Smart Innovation, Systems and Technologies

Volume 137

Series editors

Robert James Howlett, Bournemouth University and KES International, Shoreham-by-sea, UK
Lakhmi C. Jain, Faculty of Engineering and Information Technology, Centre for Artificial Intelligence, University of Technology Sydney, Sydney, NSW, Australia

The Smart Innovation, Systems and Technologies book series encompasses the topics of knowledge, intelligence, innovation and sustainability. The aim of the series is to make available a platform for the publication of books on all aspects of single and multi-disciplinary research on these themes in order to make the latest results available in a readily-accessible form. Volumes on interdisciplinary research combining two or more of these areas is particularly sought.

The series covers systems and paradigms that employ knowledge and intelligence in a broad sense. Its scope is systems having embedded knowledge and intelligence, which may be applied to the solution of world problems in industry, the environment and the community. It also focusses on the knowledge-transfer methodologies and innovation strategies employed to make this happen effectively. The combination of intelligent systems tools and a broad range of applications introduces a need for a synergy of disciplines from science, technology, business and the humanities. The series will include conference proceedings, edited collections, monographs, handbooks, reference books, and other relevant types of book in areas of science and technology where smart systems and technologies can offer innovative solutions.

High quality content is an essential feature for all book proposals accepted for the series. It is expected that editors of all accepted volumes will ensure that contributions are subjected to an appropriate level of reviewing process and adhere to KES quality principles.

** **Indexing: The books of this series are submitted to ISI Proceedings, EI-Compendex, SCOPUS, Google Scholar and Springerlink** **

More information about this series at http://www.springer.com/series/8767

Souvik Basu · Siuli Roy ·
Sipra Das Bit

Reliable Post Disaster Services over Smartphone Based DTN

An End-to-End Framework

Springer

Souvik Basu
Computer Application Centre
Heritage Institute of Technology
Kolkata, West Bengal, India

Siuli Roy
Department of Information Technology
Heritage Institute of Technology
Kolkata, West Bengal, India

Sipra Das Bit
Department of Computer Science
and Technology
Indian Institute of Engineering Science
and Technology, Shibpur
Howrah, West Bengal, India

ISSN 2190-3018 ISSN 2190-3026 (electronic)
Smart Innovation, Systems and Technologies
ISBN 978-981-13-6575-1 ISBN 978-981-13-6573-7 (eBook)
https://doi.org/10.1007/978-981-13-6573-7

Library of Congress Control Number: 2019931524

This Springer imprint is published by the registered company Springer Nature Singapore Pte Ltd.
The registered company address is: 152 Beach Road, #21-01/04 Gateway East, Singapore 189721, Singapore

The mediocre teacher tells. The good teacher explains. The great teacher inspires.

—William Arthur Ward

Dedicated to our teachers

Foreword

The monograph on *Reliable Post Disaster Services over Smartphone Based DTN* is timely and addresses an important topic. In recent years, natural disasters such as wildfires, floods, earthquakes, landslides, hurricanes, tornadoes, tsunamis and volcanoes have impacted human lives, assets and infrastructures more than ever. Indeed, natural disasters have grown almost threefold between 1980 and 2017 compared to between 1940 and 1980. Natural disasters have also affected about 217 million people each year since 1990. It has been observed that India, China, Philippines, Nepal, Bangladesh and Thailand are among the most affected, disaster-prone countries since 1995. In terms of devastation, during April 2015 Nepal earthquake, nearly 9000 people were killed, more than 600,000 houses destroyed and more than 288,255 houses damaged. Devastating floods and landslides for multiple days in June 2013 in the North Indian state of Uttarakhand not only took away more than 5700 lives but also wiped out a huge number of localities, forcing hundreds of thousands of people homeless and causing severe infrastructure damages, such as roads, bridges, power lines, electricity poles and so on.

In the aftermath of large-scale calamities, among other losses and damages, severe disruption of communication infrastructures, particularly impairment of wireless cellular towers and Internet services, is a common phenomenon. Therefore, rapid deployment of post-disaster communication infrastructure, albeit ad hoc and temporary, is crucial without which it becomes extremely challenging, if not impossible, to facilitate life-saving rescue operations as well as adequate relief services in a timely manner. The goal is to establish reliable and seamless information exchange among all stakeholders—survivors, emergency and disaster management personnel, policemen, rescue and medical teams, volunteers, government and non-government organizations, etc. This well-motivated monograph addresses some of these challenges by proposing a framework for reliable end-to-end post-disaster services by exploiting smartphone-based delay-tolerant networks (DTNs) technology that can operate despite the breakdown of traditional communication infrastructure networks.

The authors have done a good job in exploring relevant issues and challenges related to post-disaster services by proposing models, protocols and solution methodologies for accurate assessment and accumulation of various service needs, efficient resource allocation and reliable data dissemination in the wake of natural disasters. The monograph is organized into six chapters. Chapter 1 introduces post-disaster management services and communication followed by DTN architecture, routing and security issues. Chapter 2 describes various methods for post-disaster need assessment and forecasting, while Chap. 3 deals with the analysis of need accumulation over DTNs. Chapter 4 presents resource planning with the help of case-based reasoning and utility-based optimal resource allocation model. Chapter 5 investigates reliable data dissemination over DTNs using multilayer hashed encryption and group-based distributed authentication, followed by secure routing protocols. The final chapter summarizes future research directions in wireless technologies applied to post-disaster management. The authors have provided data sets from real disaster scenarios, where possible, to allow researchers to validate smartphone-based DTN protocols for post-disaster services. The proposed framework, solutions and smartphone apps may help enhance emergency preparedness and personnel training. Thus, the monograph is expected to benefit not only researchers but also disaster management practitioners of government and non-government organizations, particularly in developing countries.

Rolla, USA
December 2018

Sajal K. Das
Professor and Daniel St. Clair Chair of Computer Science
Missouri University of Science and Technology

Preface

According to the World Disasters Report, when disaster strikes, access to information is as important as access to food and water. Such information, like the status of victims and the requirement of emergency resources, is crucial for rendering timely and effective disaster relief. However, the availability of cellular and Internet connectivity in a post-disaster scenario is usually ruled out due to the failure of the supporting infrastructure. Therefore, the possibility of exchanging situational information using normal communication infrastructure gets almost restricted resulting in high perception deviation at the control station about the actual needs of the remote shelters. Under such circumstances, providing post-disaster services like *need assessment*, *need accumulation* and *resource planning* becomes truly challenging leading to futile disaster relief endeavours. Moreover, the presence of malicious entities in the network may affect the reliability of data, i.e. resource needs, at the control station. Thus, *reliable data dissemination* is another service that is to be rendered for effective disaster management.

This book deals with the problems in providing reliable post-disaster services in a challenged network environment. In particular, it proposes a framework for reliable post-disaster services over smartphone-based DTN. The components of this framework are the four services, namely need assessment, need accumulation, resource planning and finally reliable data dissemination.

The introductory chapter, i.e. Chap. 1, is dedicated to the background necessary to understand the remaining chapters of this book. It first gives an introduction to disaster management, its phases and the various post-disaster services. Next, it presents an overview of delay-tolerant networks (DTNs)—its features, architecture, routing and applications, with an emphasis on DTN security.

A principal component regression (PCR) model is derived for forecasting the exact demand for different emergency resources in a shelter, based on the dynamic influencing situational parameters, in Chap. 2. The derived model can be used to periodically forecast the ever-changing demand for different emergency resources at shelters with high precision. Next, a utility function is proposed that dynamically and quantitatively enumerates the utility of each emergency resource at different shelters. The demand forecast model along with the utility function assists in accurate *need assessment* for emergency resources at different shelters.

Assuming an underlying DTN-leveraged smartphone (carried by volunteers and relief workers)-based communication network, Chap. 3 proposes an opportunistic knowledge sharing scheme for transmitting the assessed resource needs to the control station in the absence of normal communication facilities. The knowledge sharing scheme runs on the smartphones to collect and exchange emergency resource needs at different shelters to create a consistent global view of accumulated resource needs across the entire affected area. Thus, the scheme enables *need accumulation* which in turn reduces the gap in perception about the actual needs at different shelters.

A case-based reasoning (CBR)-driven need validation technique is proposed in Chap. 4, to validate the opportunistically received resource needs at the control station. Additionally, a utility-based integer programming model for optimal resource allocation is formulated using the opportunistically transmitted and CBR validated resource needs. The model minimizes the overall resource deficit and the total resource deployment time. The CBR-driven need validation technique along with the integer programming model brings about efficient *resource planning*.

Chapter 5 proposes a series of security protocols to meet the stringent security requirements of a DTN. First, a multilayer encryption and hashing-based scheme are proposed to protect the transmitted emergency resource needs against authenticity, eavesdropping and modification attacks. Subsequently, a security-aided and group encounter-based PRoPHET routing protocol is proposed, where security features are incorporated into PRoPHET to provide security against eavesdropping, blackhole, identity spoofing and bundle store overflow attacks. Finally, a trust-based Watchdog technique is proposed and seamlessly integrated with PRoPHET to provide security against packet drop, badmouthing and ballotstuffing attacks. These security protocols run on the smartphones, carried by the volunteers and contributed towards *reliable data dissemination* so that all other post-disaster services can be rendered reliably.

There is a tremendous scope for the research community to identify novel and challenging problems in the area of ICT-based disaster management and investigate innovative solution methodologies. This book demonstrates the usage of wireless technologies coupled with smart devices to improve disaster management endeavours, by improving communication and resource planning during the aftermath of a disaster. The solutions provided herein exploit, in particular the opportunity of fast penetration of smart devices in daily life in developing countries like India, and rightly use these hand-held devices in forming delay-tolerant networks. Research conducted in this book fosters interdisciplinary collaboration among engineers, industry professionals, scientists and disaster relief organizations for the effective handling of disasters in developing countries. On the other hand, different academic regulatory bodies of such countries have made courses on disaster management mandatory at graduate levels. This book can be used as a first-hand reference for these courses. Thus, the book serves the requirements of both research and teaching communities.

Kolkata, India Souvik Basu
Kolkata, India Siuli Roy
Howrah, India Sipra Das Bit
September 2018

Acknowledgements

Publication of a book involves efforts from a large team of people apart from the authors. We would like to take this opportunity to acknowledge those who helped us at various phases of research and writing this monograph.

We express our heartfelt gratitude to Information Technology Research Academy (ITRA), Digital India Corporation, Government of India. Most of the researches presented in this monograph have been conducted as part of the ITRA-sponsored R&D project on Post-Disaster Situation Analysis and Resource Management using Delay-Tolerant Peer-to-Peer Wireless Networks (DiSARM). We would like to thank all our mentors and fellow team members of DiSARM project at other collaborating institutes, namely IIM Calcutta, IIT Kharagpur, NIT Durgapur and Kalyani Government Engineering College, Kolkata.

We are deeply indebted to Prof. Somprakash Bandyopadhyay, IIM Calcutta, who had been our mentor throughout the period of DiSARM project. His invaluable guidance enabled us to conceptualize many of the works presented in this monograph. Our brainstorming sessions always led to fresh new ideas. His constant support and inspiration have been exemplary all through our journey.

This monograph would not have been possible without the continuous support from the people who took care of all behind the scene issues, thereby allowing us to concentrate solely on preparing the manuscript. In particular, we thank Mr. Aninda Bose, Senior Editor—Hard Sciences, Springer; Ms. Shilpa, Editorial Assistant—Applied Sciences and Engineering, Springer; Mr. Ramesh Kumaran, Project Coordinator—Books Production, Springer; and Ms. Jayanthi Narayanaswamy, Project Manager, Scientific Publishing Services.

Finally, we express our gratitude to IEEE, Elsevier, ACM and Springer for allowing us to incorporate portions of our previously published works in this monograph.

Kolkata, India
December 2018

Souvik Basu
Siuli Roy
Sipra Das Bit

About the Book

The objective of the book is to present a framework for reliable end-to-end post-disaster services over smartphone-based delay-tolerant networks (DTNs), which can operate even in the absence of conventional network connectivity. History has shown that disasters like Chennai flood and Uttarakhand flash flood in India, earthquake in Nepal and tsunami in Japan severely impair Internet-based communication systems, jeopardizing people's lives—loss of life and property. Under such circumstances, providing essential post-disaster services becomes truly challenging. The book explores a number of different aspects of this challenge, ranging from accurate need assessment, to timely need accumulation, to efficient resource allocation and reliable data dissemination and presents strategies to deal with challenges. It demonstrates the usage of wireless technologies coupled with smart devices to improve disaster management endeavours, by improving communication and resource planning during the aftermath of a disaster. The primary audience of the book is technologists and researchers working in the domain of ICT-based disaster management who would find it useful in understanding the challenges of providing post-disaster services in a challenged network scenario and their possible solutions. Disaster management authorities, government agencies, NGOs and other stakeholders would also find it useful for enhancing the preparedness of volunteers and relief agencies through intelligent use of wireless technologies coupled with smart devices with the goal of better disaster management. Furthermore, the book can be used as a first-hand reference for courses on ICT-based disaster management at graduate and undergraduate levels.

Contents

About the Authors

Dr. Souvik Basu is an Assistant Professor of the Computer Application Centre at the Heritage Institute of Technology, Kolkata. He completed his Ph.D. in Computer Science & Technology at the Indian Institute of Engineering Science and Technology, Shibpur. He also holds an M.Tech. in Computer Science & Engineering from the Maulana Abul Kalam Azad University of Technology, Kolkata; an M.C.A. from the University of Burdwan; and a B.Sc. in Statistics from the University of Calcutta. With over 16 years of teaching experience, his major research interests are in opportunistic and delay-tolerant networks, network & information security, mathematical models for distributed resource management, and mobile and wireless technologies for ICT-based social applications. Dr. Basu has been the Co-Principal Investigator of a Government-funded research project on Disaster Management.

Prof. Siuli Roy is a Professor of the Department of Information Technology at the Heritage Institute of Technology. She completed her M.C.A. and Ph.D. in Computer Science & Engineering at Jadavpur University. She has more than 20 years of experience in academic research, business and software development at several organizations of international repute, including the Advanced Telecommunications Research Institute, Japan. With a unique combination of business & entrepreneurial acumen and research excellence in the area of pervasive computing and communication system, she co-founded PervCom Consulting Private Limited in 2005. Prof. Roy has authored more than 40 research publications in international journals and conference proceedings, and co-authored the book "Enhancing Performance of Ad Hoc Wireless Networks with Smart antennas", CRC Press in 2005. She has handled several Government-funded research projects as Principal/Co-Principal investigator, including projects on Disaster Management.

Prof. Sipra Das Bit is a Professor of the Department of Computer Science and Technology at the Indian Institute of Engineering Science and Technology, Shibpur. She completed her M.E. and Ph.D. at Jadavpur University. With more than 28 years of experience in academics and research, her major interests are in delay-tolerant networks, wireless sensor networks, mobile computing, low-overhead multimedia processing over wireless networks, and low-overhead wireless network security. She has published extensively in these areas. Prof. Das Bit has handled several Government-funded research projects as Principal/Co-Principal investigator, including projects on Disaster Management.

Abbreviations

ACF	Action Contre La Faim
AES	Advanced Encryption Standard
AF	Accuracy factor
ANN	Artificial neural network
ARD	Average resource deficit
BP	Back-propagation
BPNN	Back-propagation neural networks
CBR	Case-based reasoning
CID	Control node ID
CN	Control node
CPin	Control pin
CRL	Certificate Revocation List
DDH	Decisional Diffie–Hellman
DFY	Doctors For You
DOM	Disaster operations management
DoS	Denial of service
DP	Delivery predictability
DTN	Delay-tolerant network
EMD	Expected message delay
FC	Forwarding competency
FE	Forwarding evidence
FEMA	Federal Emergency Management Agency
FIFO	First in first out
FN	Forwarder node
GID	Group ID
GP	Group pin
GPRS	General Packet Radio Service
GSM	Global System for Mobile
IBC	Identity-based cryptography
IBE	Identity-based encryption

ION	Interplanetary Overlay Network
ISP	Influencing situational parameter
ITRM	Iterative Trust and Reputation Mechanism
LEO	Low Earth orbiting
MANET	Mobile ad hoc network
MCPin	Modified control pin
MGP	Modified group pin
MLCE	Multiple layer commutative encryption
MLR	Multiple linear regression
NSE	Nash–Sutcliffe efficiency
OppNet	Opportunistic network
PB	Percentage bias
PC	Principal component
PCA	Principal component analysis
PCR	Principal component regression
PCRA	Principal component regression analysis
PD	Perception deviation
PEKS	Public encryption with keyword search
PEON	Privacy-Enhanced Opportunistic Networking
PFM	Positive feedback message
PGP	Pretty Good Privacy
PKI	Public key infrastructure
PRoPHET	Probabilistic Routing Protocol using History of Encounters and Transitivity
QRC	Qatar Red Crescent
RMSE	Root mean squared error
RPGM	Reference point group mobility
RSA	Rivest–Shamir–Adleman
SA	Situational awareness
SAGE-PRoPHET	Security-aided and group encounter-based PRoPHET
SHA	Secure Hash Algorithm
SN	Shelter node
SnW	Spray and wait
SRD	Shelter-specific resource deficit
TA	Trusted authority
TCP/IP	Transmission Control Protocol/Internet Protocol
T-PRoPHET	Trusted PRoPHET
TTP	Trusted third party
VANET	Vehicular ad hoc network
VDTN	Vehicular DTN
Wise-PRoPHET	Watchdog supervised PRoPHET
WLAN	Wireless local area network

Chapter 1
Introduction

Large-scale natural disasters like tsunami, earthquake, cyclone, etc. have serious impact on the telecommunication infrastructure. Cellular network/Internet services immediately become non-functional or partially functional in emergencies due to the failure of the supporting infrastructure through system damage [1]. Therefore, the possibility of information exchange using normal communication infrastructure is almost ruled out. However, any emergency response activity consists of several interrelated phases each of which requires appropriate situational information for its execution. The gap between the need and availability of crucial situational information poses severe challenges in providing effective disaster management services to the affected community. This book addresses these challenges and proposes a framework for reliable post-disaster services in a challenged network environment.

1.1 Disaster Management

As per the Oxford dictionary, a disaster is "a sudden accident or a natural catastrophe that causes great damage or loss of life" [2]. In other words, disaster is a sudden, calamitous event that seriously disrupts the functioning of a community or society and causes human, material, and economic or environmental losses that exceed the community's or society's ability to cope using its own resources [3]. Disaster management, according to the Disaster Management Act (No. 53) 2005 [4], is a continuous and integrated process of planning, organizing, coordinating and implementing measures which are necessary and expedient for prompt response to any threatening disaster situation. It is the organization and management of resources and responsibilities for dealing with all humanitarian aspects of emergencies, in particular preparedness, response and recovery in order to lessen the impact of disasters [5].

S. Basu et al., *Reliable Post Disaster Services over Smartphone Based DTN*, Smart Innovation, Systems and Technologies 137,
https://doi.org/10.1007/978-981-13-6573-7_1

1.1.1 Disaster Management Phases

Disaster management is a phased sequence of activities, also termed as the disaster management cycle. The three key phases of disaster management are pre-disaster phase, during disaster phase and post-disaster phase [6]. These three phases encompass the six main activities pertaining to disaster management, namely—*prevention, mitigation, preparedness, response, recovery* and *development*.

- ***Pre-disaster Phase***: Also called the risk reduction phase reduces the potential for human, material or environmental losses caused by hazards to ensure that these losses are minimized when the disaster actually strikes. There are three main activities in the pre-disaster phase—*prevention, mitigation, preparedness*.
- ***During Disaster Phase***: Ensures that the needs and provisions of victims are met to alleviate and minimize suffering.
- ***Post-disaster Phase***: Also called the recovery phase, attempts to achieve rapid and durable recovery that does not reproduce the original vulnerable conditions. There are three main activities in the post-disaster phase—*response, recovery and development*. Response refers to the activities immediately after any calamity, which includes restoration of communication, setting up shelters and control rooms, evacuation and rendering medical aid to the needy. Immediate damage and need assessment are undertaken within this activity. This also includes supplying emergency resources like food, drinking water, clothing etc., to the affected community and disbursement of assistance in cash or kind. Recovery refers to the activities involved in bringing the affected area back to some degree of normalcy. Development attempts to return the affected communities to improved pre-disaster functioning leading to social continuity. Figure 1.1 [6] shows the typical disaster management cycle comprising three phases and six activities.

1.1.2 Post-disaster Services

One of the crucial activities in the post-disaster phase is distribution of emergency resources to the affected population to reduce the impact of the disaster. Equitable distribution of emergency resources among the victims, conforming to some global policy, requires certain post-disaster services to be offered. Extensive literature survey is conducted to identify the post-disaster services that play pivotal role in effective resource distribution. According to the Disaster Assistance Guide to Recovery Programs, FEMA [7], National Post-Disaster Recovery Planning and Coordination, UNDP [8], Disaster Services Start-up Guide, Corporation for National and Community Service [9], post-disaster Needs Assessment, International Recovery Platform [10], three essential post-disaster services are—*need assessment*, *need accumulation* and *resource planning*. The services along with the challenges in rendering them are discussed below.

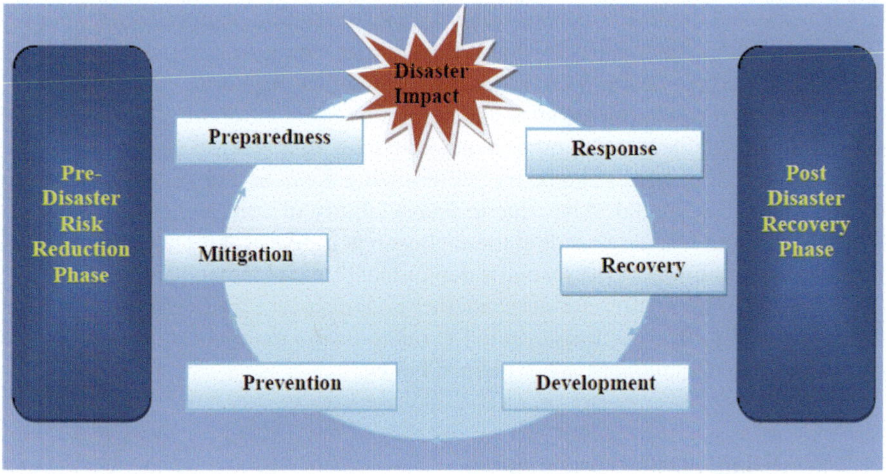

Fig. 1.1 Disaster management cycle [6]

1.1.2.1 Need Assessment

Post-disaster need assessment comprises an approach to harmonize the assessment, analysis and prioritization of needs by a range of stakeholders [10]. It refers to the assessment of exact need (i.e., demand and utility) for emergency resources at the shelters. Forecasting the exact demand for emergency resources in a relief shelter is challenging owing to the fact that situational parameters that influencing these demands are constantly changing with the varying dynamics of the disaster. Furthermore, measuring the correct utility of such resources is tricky, where demands are constantly surfacing and supplies drop in at an non-uniform rate.

1.1.2.2 Need Accumulation

This refers to the collection and transmission of shelter needs to the control station for necessary resource allocation. However, since cellular and other communication facilities typically become fragile or non-functional during disasters, it becomes difficult to collect the resource needs from remote and inaccessible shelters and transmitting them to the control station. This results in a gap in perception at the control station about the actual need for resources at the shelters.

1.1.2.3 Resource Planning

Resource planning is a plan for using available resources, especially in the near term, to achieve goals for the future. It is the process of allocating scarce resources among the various stakeholders [7, 9]. This refers to the formulation of an effective resource allocation strategy. On one hand, needs transmitted over a frail network are highly susceptible to corruption and unavailability. Using this inaccurate and partial need information for formulating resource allocation strategies, the entire allocation would become unreasonable. On the other hand, resources in a post-disaster scenario are scarce and not all demands can be met; minimizing the deficit in allocation of high-utility resources is crucial. Also, due to unavailability of communication facilities, information about resource needs at the shelters reach significantly late at the control station. Thus, minimizing the resource deployment time is inevitable.

In addition to these three services, another post-disaster service that plays pivotal role in providing effective disaster relief is *reliable data dissemination*. The service and its associated challenges are explained below.

1.1.2.4 Reliable Data Dissemination

This refers to gathering reliable resource needs. The success of the resource planning service depends upon the reliability of the situational data, in terms of shelter resource needs, transmitted over the network. However, malicious and unscrupulous entities may launch several attacks on the network, jeopardizing the reliability of the transmitted resource needs. This affects the delivery, accuracy and timeliness of resource needs at the control station, leading to delayed and incorrect resource planning. In the absence of the stable communication infrastructure and trusted third parties in a post-disaster scenario, implementing traditional security mechanisms like cryptographic signatures, certificates and authentication protocols becomes impossible.

1.1.3 Post-disaster Communication

Each of the tasks in the post-disaster phase, discussed above, requires appropriate situational information for getting executed. First responders, rescue workers and volunteers deployed in the affected area need comprehensive, reliable and high quality situational information in order to conduct accurate need assessments and effective resource planning. It, is therefore, important to enable data collection from inside a disaster area and then delivering it to a service point where it can be processed, actuated and shared [11]. Nevertheless, disasters hinder usual communication experiences of people because traditional communication options, e.g., cellular communication or Internet, may not be available after a disaster due to infrastructure damage or power outage [1]. This calls for alternative approaches to meet the communication needs until the normal communication infrastructure is operational again.

In recent years, there has been an enormous propagation of wireless devices, such as smartphones, with various connectivity options, such as Wi-Fi Direct and Bluetooth. This makes communication experiences quite pervasive, even when people are moving. These devices with their connectivity capabilities can be exploited to form a delay-tolerant network (DTN) in a post-disaster scenario [11]. Researchers working in this domain have strongly proposed setting up an emergency post-disaster communication network using smartphones working in the DTN mode [12–16]. Other low-delay alternatives, such as satellite communication, are expensive and require specialized devices that may not be readily available to volunteers and relief workers. Android-based smartphones running in the ad hoc mode are the most widely used smartphones for DTN implementations. For enabling ad hoc mode in android phones, CyanogenMod, an open-source (after-market) distribution based on the Android Open Source Project (AOSP) is used [13]. However, CyanogenMod, like any other firmware, comes with (some) flaws and has room for further improvements. Carbon Rom, Slim Rom and MIUI may be used as alternatives to CyanogenMod.

1.2 Overview of Delay-Tolerant Networks

The Internet today widely operates on TCP/IP, thereby ensuring a standard set of protocols over which millions of devices worldwide operate. However, these protocols may not be applicable to all kinds of networks, particularly those that operate under the constraints of high delays and losses, for example, terrestrial wireless networks, military networks and atypical media networks. These networks require protocols that take into consideration the specific communication needs like link connectivity, delay, data rate asymmetry, addressing, reliability mechanisms and quality of service [17]. Delay-tolerant network (DTN) poses an amazing potential of handling these requirements of specialized networks, especially over long distances. DTN has been a topic of interest in networking for regions where constant back-to-back connectivity is not available [17, 18]. A DTN is a system of provincial systems which is demonstrated to be an overlay over the present existing systems including the web. DTN bolsters interoperability among an arrangement of fluctuated networks by allowing long interruptions and postponements between and inside those systems [19–21]. In spite of the fact that DTNs were initially created for interplanetary utilize, where delay-resilience is the best need, they have unquestionable applications on earth [19, 22]. Figure 1.2 [18] shows a typical DTN that connects the Internet, various networks on earth and subnets across the solar system.

1.2.1 Features

Compared to traditional Internet, mobile ad hoc networks and WLANs, DTNs have a unique set of features [21] that are described below:

- ***Intermittent Connectivity***: The lack of continuous path between source and destination and the mobility of nodes create frequent network partitioning in a DTN leading to intermittent connectivity.

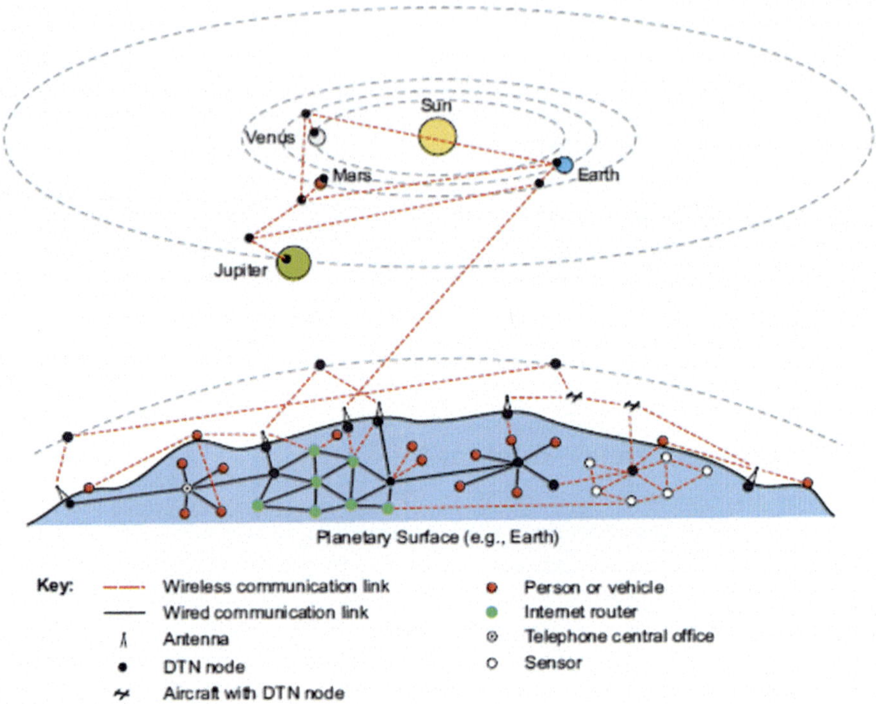

Fig. 1.2 A typical DTN [18]

- ***High Delay***: Notwithstanding discontinuous availability, long engendering deferrals among nodes and variable queuing delays at nodes add to end-to-end delays.
- ***Asymmetric Data Rates***: Intermittent connectivity and high end-to-end delay leads to low and asymmetric up-down link data rate.
- ***Dynamic Topology***: Environmental changes, energy depletion or other failures, which results in dropping out of the network, make the DTN topology extremely dynamic.
- ***Heterogeneous Interconnection***: DTN is an overlay network that supports interoperability among a host of heterogeneous networks for transmission of asynchronous message.

1.2.2 Architecture

Being an overlay network, DTN is proposed to work over the current protocol stacks in different network structures and offer a store-and-forward entryway between them when a node physically exists in at least two divergent networks. Interoperability between these networks is accomplished by special DTN gateways located at their

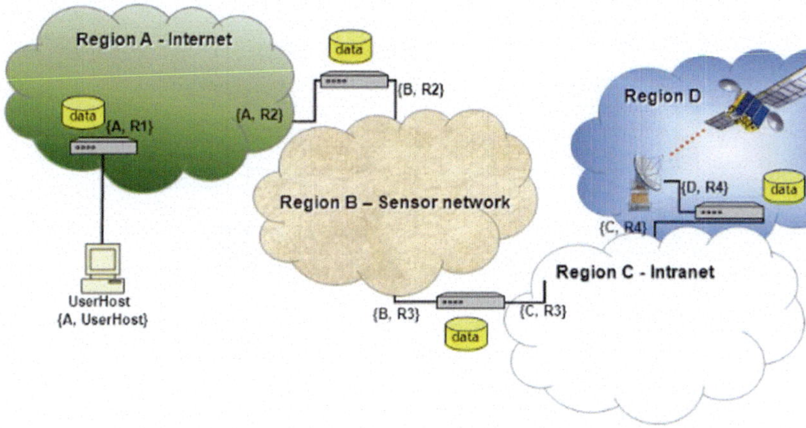

Fig. 1.3 DTN architecture [18]

interconnection points. The DTN engineering [17, 18] incorporates the ideas of regions and DTN gateways, as shown in Fig. 1.3 [18]. In this example, four regions are outlined (A, B, C, D). Region B incorporates a DTN gateway occupant on a passenger transport that cycles between DTN gateways 3 and 5. Region D incorporates a low earth orbiting satellite connection (LEO) that likewise gives occasional network (though maybe more normal than the transport which might be liable to vehicular blockage or different postponements).

Region limits are utilized as interconnection points between different network protocol and addressing families. All the more formally, two nodes are in a similar region if they communicate without utilizing DTN gateways (for the most part utilizing existing protocols local to the containing region). Few region types (e.g. Internet-like, ad hoc mobile and sporadically connected) might build up and each instance of the parallel kind will realize an analogous stack of underlying protocols. DTN gateways keep up a correspondence to the waypoint concept that illustrates a point through which data should pass so as to gain entry to a region. This point can provide as a foundation for both translations (between region-specific encodings) and a point to implement policy and control. In working above the transport layer, nevertheless, DTN gateways are focused on reliable message routing instead of best-effort packet switching. These gateways are in charge of storing messages in non-volatile storage when trustworthy delivery is necessary and mapping between conflicting transports by resolving globally significant name tuples to locally resolvable names for traffic destined internally to an adjacent region. They may also carry out authentication and access control tests on incoming traffic to ensure forwarding is to bc allowed. Message forwarding is implemented using the bundle protocol discussed below.

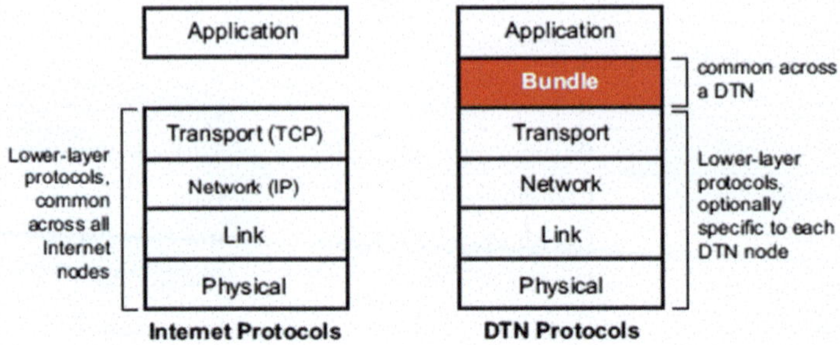

Fig. 1.4 Bundle layer [18]

1.2.2.1 Bundle Protocol

The DTN architecture implements store-and-forward message switching by overlaying a new transmission protocol, called the bundle protocol [19], on top of lower-layer protocols, such as the Internet protocols. This protocol defines a series of contiguous data blocks as a bundle—where each bundle contains enough semantic information to run the application. Bundles are routed in a store-and-forward manner between participating nodes over varied network transport technologies (including both IP and non-IP-based transports). The bundle protocol binds the lower-layer protocols in order that application programs can communicate across similar or dissimilar sets of lower-layer protocols under long network delays or interruptions. The bundle protocol agent stores and forwards whole bundles (or bundle portions) between nodes. A single bundle protocol is employed all over a DTN. The lower-layer protocols under the bundle protocol are selected to go with the features of each communication setting. Figure 1.4 [18] exemplifies the bundle protocol overlay while contrasting the Internet protocol stack with a DTN protocol stack.

1.2.3 Routing

Routing is a key feature in any kind of network including DTNs. DTN routing protocols can be broadly categorized as single copy protocols and multicopy protocols. In single-copy protocols, messages are not replicated. Once a message is forwarded to an encountered node, it is removed from the buffer of the sender. Multicopy protocols are broadly categorized into—limited and unlimited multicopy routing. In the limited version, a message is replicated for a certain number of times and forwarded to several nodes. In the unlimited version, there is no set upper bound on the number of replications of a given message. An extreme case is flooding is the case where a message is replicated and forwarded to all the nodes in the network [22, 23]. In this section, a few classical [24–28] as well as new-generation [29–32] DTN routing protocols are briefly explained.

- ***Direct Delivery and First Contact*** [24]: These are single-copy routing protocols. Here, just one copy of every message is created in the network. In Direct Delivery, the source node delivers its messages to the destination node(s). On the contrary, in First Contact a node forwards a message to the first node that it encounters, and the process continues until the message reaches its destination. Thus, messages may be handled by several nodes. Both protocols use FIFO queuing. Single-copy routing is uncommon in DTNs, since intermittent connectivity results in low delivery ratio of the messages as well as high delivery latencies. These schemes are hardly used in practice.
- ***Epidemic*** [25]: In this protocol, messages are broadcast to all nodes that are encountered. When the message buffer is full, oldest messages get deleted. Messages are forwarded according to a FIFO policy, and there is no upper bound on the number of replicas of a message.
- ***PRoPHET*** [26]: In this protocol, a metric called delivery predictability is used to evaluate the probability of a node reaching the destination. Delivery predictability is updated when a new node is encountered or the time out period expires. A node delivers a message to all encountered nodes that have a higher delivery predictability than its own. PRoPHET has similar queuing and replication techniques as Epidemic (i.e., FIFO and unlimited), however, the forwarding is based on delivery predictability and is thus history based.
- ***Spray and Wait*** [27]: In this protocol, there are two phases. In the spray phase (executed only once), M copies of the message are primarily percolated to M distinct "relays". In the wait phase, if the destination is not reached in the spray phase, the M nodes carrying a copy of the message execute direct transmission. The forwarding technique in this protocol is a mix of always (Spray) and direct delivery (Wait). The number of replication of the message is limited by M and queuing is FIFO.
- ***MaxProp*** [28]: In this routing protocol, every node maintains a routing table, which describes the probability (i.e., cost) to get to another node via its present neighbours. Routing tables are updated according to the information received from the neighbours, and messages are ordered and forwarded according to their cost to arrive the destination. Here, forwarding is history based, and replication is limitless. The Dijkstra algorithm is used to evaluate the cost to reach the destination and the current number of hops that a message has accumulated.
- ***RAPID*** [29]: In RAPID, messages are prioritized as per a utility that depicts the expected delay to reach the destination. Here DTN routing is treated as a resource allocation problem. The protocol interprets routing metrics as per-message utilities, which decide how messages are to be replicated in the network. Messages with lower utility are deleted to accommodate new messages. The forwarding technique is history based where messages with high marginal utility are forwarded first. Here, queue management is destination dependent and replication is unlimited.
- ***SimBet*** [30] ***and BubbleRap*** [31]: These are both social-based protocols. In SimBet, similarity and betweenness functions are pooled to choose the message to be forwarded, and there is a single copy of the message in the network, implying zero

replication. BubbleRap mixes the information on the community structure and the centrality of the nodes to make a decision whether to forward. Each message has two kinds of ranking, namely—global and local, related to its community. Message forwarding is made by the global ranking, until the message arrives at a node that is in the same community of the destination. At this point, local ranking is used until the destination is reached or the message perishes. Nodes, forwarding a message, do not delete their copies until the message is delivered to the community of its destination. In this protocol, forwarding is social based and replication is unlimited.
- ***Encounter-Based Routing*** [32]: This routing brings in the idea of quota-based replications, where the number of duplications of a message is not dependent on the number of nodes in the network. In this protocol, each node stores their average encounter rate with the other nodes by applying exponentially weighted moving average method. Each node stores the node's past rate of encounters in a variable and the number of encounters with the node that happened in the current time window in another variable.

1.2.4 Applications

DTNs can be used as a possible mode of communication in several challenged environments like deep-space and underwater sensor networks. A few known realizations of DTN [22] include the DTN2 project, Interplanetary Overlay Network (ION), Postellation, and IBR-DTN. Some real-world applications of DTN include wildlife tracking, underwater sensor networks, post-disaster communication networks, networks for remote areas or rural areas in developing countries, vehicular networks and pocket-switched networks [19, 22]. DTNs are vulnerable to irregular connectivity and disconnection of nodes because of constraints in power supply, node mobility, meagre node density, and equipment malfunctions. As discussed in Sect. 1.1.3, the researchers working in the networking domain have strongly patronized the use DTN for establishing emergency post-disaster communication networks through DTN enabled smartphones [12–16]. Other low-delay alternatives, such as satellite communication, are expensive and require specialized devices that may not be readily available to disaster relief volunteers and rescue workers.

In the following sections, first a few DTN-based systems that provide emergency communication facilities in a post-disaster scenario are discussed. Subsequently, a few other DTN-based systems that provide communication solutions which are not emergency in nature are discussed.

1.2.4.1 Systems for Emergency Communication

Kevin Fall et al. present a disruption-tolerant system for secure and efficient disaster response communication [12]. This system describes a situational awareness

(SA) application architecture for emergency scenario when the network infrastructure is degraded or barely functioning. The architecture comprises back-end server and portable nodes (producer/consumer devices) to collect and access data. It also comprises field servers that are set up to act as ad hoc routers and caches for data. Back-end servers are typical servers found in an Internet data centre. The SA service uses a combination of DTN and Internet protocols.

Hossmann et al. develop Twimight [14], a Twitter application relying on DTN communications to spread tweets and sensor data in an epidemic fashion. It is an open-source Twitter client for android phones featured with a disaster mode, which users enable upon losing connectivity. In the disaster mode, tweets are not sent to the online Twitter servers but stored on the phone, carried around as people move, and forwarded opportunistically via Bluetooth or Wi-Fi Direct when in proximity with other smartphones.

Mehendale et al. propose LifeNet [15], a flexible Wi-Fi-based data communication solution designed for post-disaster scenarios. It is open-source software and designed to run on consumer devices such as laptops, smartphones and wireless routers. LifeNet is an ad hoc networking platform over which critical software applications including chat, voice messaging, MIS systems, etc. can be easily installed. It can grow incrementally and is robust to node failures and enables Internet sharing. A novel multipath ad hoc routing protocol present at its core enables LifeNet to achieve these features.

Ntareme et al. propose Bytewalla [16], an Android-based system to enable DTN. Bytewalla uses the DTN version 2 specification by the DTN research group to send and receive data bundles. A user can indicate where to accumulate the bundles along with how much memory to be used. It has facilities for sending emails via DTN, which is beneficial for rural areas with no Internet facility. A DTN server receiving such emails translates them into bundles and forwards them to smartphones using the application. It also takes out the emails from bundles before delivering it to a client.

1.2.5 Other Communication Systems

DakNet [33], developed at MIT Media Lab, is a DTN-based system to offer cheap and energy-conserving communication services. It combines asynchronous services like email and file transfers along with wireless communication to hook up remote villages with towns and cities. It consists of kiosks, mobile access points (MAPs) having portable storage devices and Internet access points or hubs. The MAPs are attached with and powered by different types of vehicles. Whenever such a MAP comes in contact with a kiosk, data are exchanged between them. Subsequently, when a MAP comes in proximity to a hub (in town or city), data from the kiosks are synched with the Internet. MAPs, therefore, play the role of data mules. DakNet has been widely used in villages of India and Cambodia.

Seth et al. in [34] have presented an architecture for providing the Internet service in rural areas using delay-tolerant network. Ferries (or Mechanical Backhauls) upload and download data opportunistically to and from an Internet gateway, which is a computer that has a Wi-Fi interface, storage, and an always-on connection to the Internet. The ferry is just a mobile DTN router and Internet gateways and kiosk controllers are examples of fixed DTN routers. When the ferry is in the range of kiosks, both the devices opportunistically transfer their data. A gateway collects data opportunistically from a ferry and stores it in local storage before uploading it to the Internet. It also downloads bundles on behalf of kiosk users and transfers them opportunistically to the appropriate ferry.

Ginzboorg et al. [35] present an architecture that can be used for data communication in mines with the usage of delay tolerant opportunistic communication network. This is a system that provides communication solution in a mine using the technique of store-carry-forward of messages by intermediary nodes, without the need for fixed infrastructure. This method of communication is facilitated by the fact that the movement of people and equipment in a mine is mostly deterministic. The personal devices held by the workers in mining regions communicate and share information relevant to various mining operations. The system also increases the level of optimization and control of the mining operation.

1.3 Security in DTN

Security is one of the major obstacles in large-scale application of DTNs. As in conventional mobile ad hoc networks, the use of open networks and multihop transmission makes DTNs susceptible to a variety of security attacks. Misbehaving nodes may attempt to gain unauthorized access to service, resource or information affecting message integrity, authenticity, confidentiality and system performance. Then again, due to the exceptional characteristics of DTNs including extended round-trip delay, recurrent network partitioning, opportunistic routing and restricted computational and storage capacity, present security protocols for the conventional ad hoc networks either fail completely or perform miserably when applied to DTN scenario. Thus, a set of fresh security protocols are extremely desirable to handle the strict security and efficiency conditions for securing DTNs.

1.3.1 Security Requirements

There are five fundamental security requirements that need to be ensured for reliable and effective functioning of a DTN. These requirements, as explained in [36, 37], are outlined below:

- ***Authentication***: As in traditional systems, authentication schemes confirm the identity of the DTN nodes and distinguish lawful DTN users from illegal users. In DTNs, it is necessary for each intermediary DTN node to have the ability to check that the message was indeed forwarded by an authorized node, at a satisfactory rate or class of service for which they are approved.
- ***Confidentiality***: Since radio channels are predominantly open to eavesdropping, the confidentiality prerequisite is to ensure that sensitive messages are not revealed to unauthorized third parties while the bundles propagate over DTN connections.
- ***Integrity***: Since radio channels are also open to active attacks, the integrity of messages must be properly preserved. Integrity prerequisite should make sure that the forwarded messages are not altered during the transmission process.
- ***Anonymity***: A network must disclose neither the location of the user nor the party with which it communicates. In some applications, ensuring identity and location anonymity is an extremely significant security requirement. Anonymity is an additional requirement and more linked to the requirements of particular DTN applications.
- ***Availability***: Illegal access and utilization of limited DTN resources are a grave apprehension in DTN security. The availability requirement ensures that network resources are available for and to authorized entities whenever required, restricting unauthorized entities from using such resources.

1.3.2 Attacks on DTN

Generally speaking, the misbehaving nodes in a DTN can be categorized as—malicious and selfish [36]. The aim of the former is to attack the correct network operations with no personal gain. Conversely, selfish nodes make the most of their own gains or collective gains with colluding nodes while minimizing their contributions to it. Attacks launched by these misbehaving nodes can be classified into three classes—Privacy Attacks, Communication Attacks and Trust/Reputation Attacks. Figure 1.5 shows a subset of possible attacks [36, 38, 39] on DTNs. Each of the attacks is elaborated below and shown in Fig. 1.5.

1.3.2.1 Privacy Attack

These types can be classified as—confidentiality, modification and authenticity attacks.

- ***Confidentiality Attack***: This is an attack on the confidentiality of transmitted information. It is responsible for eavesdropping and sniffing. In other words, the adversaries collect useful information being transmitted between the nodes. This information is used further for launching attacks that disturb the normal operation of

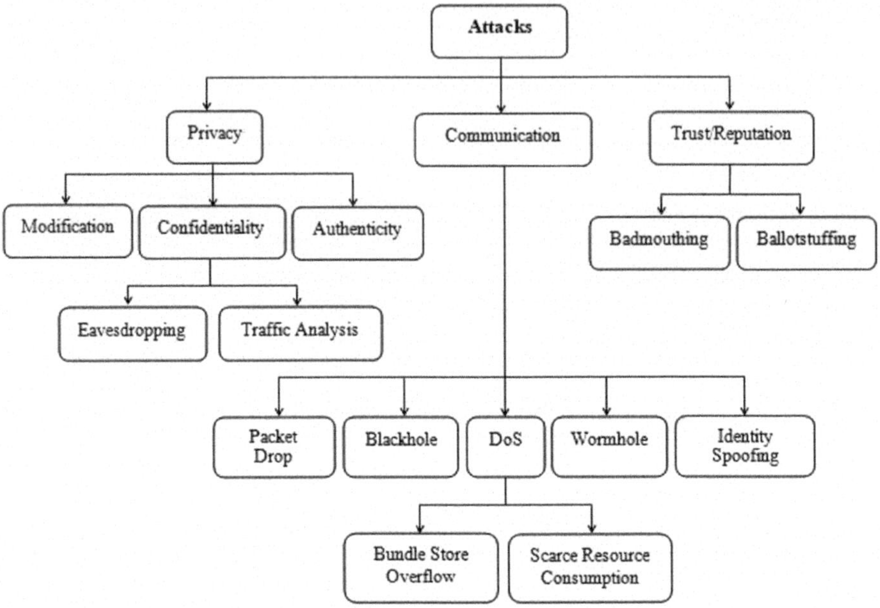

Fig. 1.5 Attacks on DTN

the network. Based on variation in techniques of getting information being transmitted, confidentiality attacks are classified as:

- **Eavesdropping Attack**—In this attack, the adversary discovers the communication contents by listening to the data being transmitted.
- **Traffic Analysis Attack**—In this attack, the traffic pattern is analysed by the adversaries. Based on this analysis, the adversaries cause harm to the network in the form of malicious activities e.g., finding out location of nodes, etc.

– ***Modification Attack***: This is an attack on the integrity of transmitted information, where messages (or bundles) travelling through heterogeneous networks are modified for malicious purposes.
– ***Authenticity Attack***: This is an attack on the authenticity of nodes, where data packets are sent by unauthorized nodes with several malicious intentions.

1.3.2.2 Communication Attack

In this type of attack, the attacker disrupts communication in the network either by blocking the frequencies or by replaying packets. Five categories of communication attacks are as follows:

- ***Denial-of-Service (DoS) Attack***: This is defined as the result of any action that prevents any part of a DTN from functioning correctly or in a timely manner. Based on the implementation techniques, there are several types of DoS attacks. Two important DoS attacks are described below:

  - **Bundle Store Overflow Attack**—In this attack, attackers introduce false DTN bundles to consume precious DTN resources. Further, malicious nodes may redistribute older bundles before it expires from known sources which worsen storage overflow if the message is replayed before it expires.
  - **Scarce Resource Consumption Attack**—In this attack, unauthorized applications control certain DTN infrastructure (e.g. by attacking a routing protocol), as a result of which, the resource consumption could be catastrophic for the networks. Due to the resource scarcity in DTNs, unauthorized access and use of DTN resources can be a serious concern.

- ***Packet Dropping Attack***: This is a case where a selfish node chooses to drop packets entirely or selectively.
- ***Black hole Attack***: This is an attack where a malicious node attracts all messages from nodes it meets and forwards none of them.
- ***Wormhole Attack***: This is an attack the attacking node captures the packets from one location and transmits them to other distant located node which distributes them locally. A wormhole attack can easily be launched by the attacker without having knowledge of the network or compromising any legitimate nodes or cryptographic mechanisms. The tunnel is either the wired link or high-frequency links. This creates the illusion that the two end points of the tunnel are very close to each other.
- ***Identity Spoofing Attack***: This attack, also known as Sybil Attack, is an attack where a malicious node impersonates other nodes or claims false identities of nodes. In the worst case, an attacker may generate an arbitrary number of node identities using only one physical device.

1.3.2.3 Trust/Reputation Attack

In this type of attack, malicious nodes manipulate the trust/reputation values of forwarding nodes by either bad-mouthing or ballot-stuffing. Detection of selfish nodes becomes extremely challenging in the presence of such attacks; a large number of messages are directed towards selfish nodes and are eventually dropped.

- ***Bad-mouthing Attack***: This ruins the reputation of well-behaved nodes. Malicious nodes set the trust value of all altruistic nodes to 0, ruining their reputation.
- ***Ballot-stuffing Attack***: This boosts the reputation of selfish nodes. Malicious nodes set the trust value of all selfish nodes to 1, boosting their reputation.

1.3.3 Challenges in DTN Security

There are certain unique security challenges [36] that distinguish DTN security from conventional network security. These challenges bring in fresh possibilities to take care of the security issues from a new perception, which is completely different from traditional network security. The challenges are discussed below:

- ***Lack of End-to-End Connectivity***: Lack of end-to-end connectivity in a DTN not only creates challenge in routing but also makes the existing security solutions unsuitable. As for example, end-to-end confidentiality using conventional encryption techniques needs multiple-round key prior agreement between the sender and the receiver. Nonetheless, in DTNs, such key agreement is infeasible considering the fact that there is no direct network connectivity at the time of sending messages. The same is true for mutual authentication. Lack of end-to-end connectivity makes public key certificate revocation challenging as well. In a conventional public key infrastructure, the most frequently used certificate revocation method is through certificate revocation list. But, in DTNs, the nodes may experience delayed or frequent loss of connectivity to certificate revocation servers.
- ***Fragmentation***: Because of high mobility in a DTN, a network connection becomes accessible only for a small period. Thus, when a message is big, it may be impossible to transmit the complete message at one go. One likely solution to this problem is to split the message into smaller parts and transmit some parts of a big message through the current link and rest of the parts using another link later to make the best use of restricted resources. Due to this fragmentation problem, traditional authentication scheme, like the signing the entire message by the sender seems infeasible, since the intermediate receiver is unable to authenticate the received fragments before receiving the entire message.
- ***Resource Scarcity***: Scarcity of resources is one more key concern in DTN security design. Due to limited contact time, DTN nodes have to receive, check and forward large number of bundles in a short period. Thus, bandwidth constraint and computational burning up are critical issues in DTN security. On one hand, security operations such as authentication are regarded as a necessity to protect precious DTN resources, in terms of bandwidth and computational power, from unlawful access and use. On the other hand, security techniques will themselves inescapably bring in additional computation and transmission overheads.
- ***Bundle Accumulation***: Because of the store-carry-forward dissemination feature, the bundles may accumulate at some intermediate nodes. An intermediate bundle forwarder may contemporarily receive, store and authenticate multiple bundles from different senders before these bundles are transmitted to the next-hop. Since authentication process usually involves computationally heavy operations such as signature verification, the collected authentication-related security operations may incur huge computational overhead, which proves conventional security solutions inappropriate for DTNs.

1.4 Overall Objective and Contribution of the Book

Appropriate situational information is crucial for each of the post-disaster services—*need assessment*, *need accumulation* and *resource planning*, discussed in Sect. 1.1.2, to be effectively rendered. First responders, rescue workers and volunteers deployed in the affected area need comprehensive, reliable and timely situational information in order to conduct accurate need assessments and effective resource planning. However, the availability of cellular and Internet connectivity in a post-disaster scenario is almost ruled out due to the failure of the involved infrastructure. So, the prospect of collecting situational information using usual communication infrastructure gets ruled out resulting in a gap in perception about the actual needs of the remote shelters at the control station. Such circumstances throw serious challenges in providing reliable post-disaster services. Moreover, malicious and unscrupulous entities may launch several attacks on the network, jeopardizing the reliability of the transmitted resource need, calling for *reliable data dissemination*. These challenges are the motivating factors for devising strategies that render reliable post-disaster services in the absence of traditional communication facilities, using alternative solutions. The overall objective of this book is to propose a framework for reliable post-disaster services over smartphone-based DTN. The components of this framework are the four services, namely—need assessment, need accumulation, resource planning and finally reliable data dissemination. Figure 1.6 shows a schematic representation of the proposed framework. The specific contributions, classified on the basis of services rendered by the shelters, control station and volunteers, are listed below.

– A principal component regression (PCR) model is proposed for predicting the precise demand for different emergency resources in a shelter, depending on the influencing situational parameters. The proposed model can be utilized to forecast

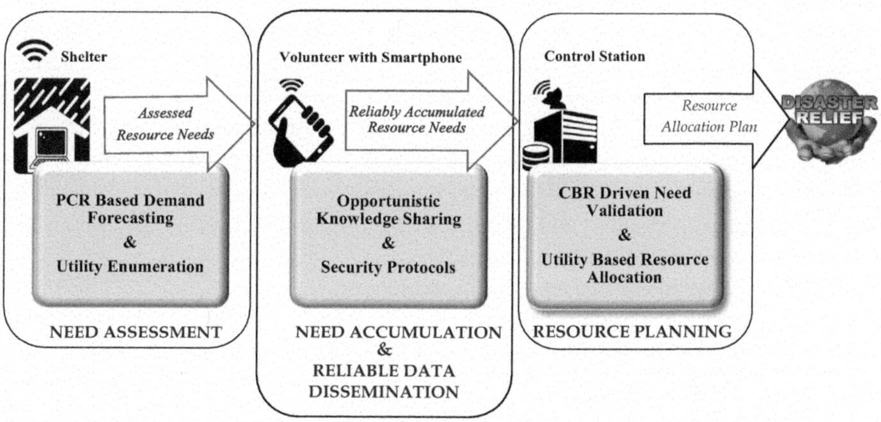

Fig. 1.6 A framework for reliable post-disaster services over smartphone-based DTN

the ever-changing demand for different emergency resources at shelters with high precision. In addition, a utility function is derived that dynamically and quantitatively measures the utility of emergency resources at different shelters. Situational parameters of a shelter like—number of victims, category and intensity of emergency they have, their resource utilization pattern, speed of resource replacement, rate of migration, etc. are uploaded to the shelter-node (a laptop or workstation in the shelter), set-up therein, by the emergency managers present on site. The shelter-node estimates the demand for different emergency resources using the proposed PCR model and measures the utility of these resources using the derived utility function. These demands and utilities are next forwarded to the control station where they are authenticated and used for resource planning. The demand forecast model and the utility function help in exact need assessment for emergency resources at different shelters. Chapter 2 handles post-disaster need assessment. Demand Forecast using principal component regression model and utility enumeration through utility function is described in this chapter.

- An opportunistic knowledge sharing scheme is designed for forwarding the shelter resource needs to the control station over a DTN-based smartphone operated communication network. The scheme executes on the smartphones (carried by the volunteers) and uses stable human mobility characteristics to gather and exchange emergency resource needs at different shelters. Subsequently, it generates a comprehensive view of resource needs across the entire affected area and forwards the same to the control station. Chapter 3 illustrates the opportunistic knowledge sharing scheme for need accumulation over DTN.
- A case-based reasoning (CBR) driven need validation to be carried out at the control station is proposed. The validation technique—(a) verifies the received resource needs, (b) rectifies the corrupted or stale needs and (c) projects the unavailable needs. Subsequently, a utility-based integer programming model is formulated for optimal resource allocation using the opportunistically transmitted and CBR validated resource needs that minimize the overall resource deficit and the total resource deployment time. The control-node, a workstation/server at the control station, receives resource needs from the forwarder-nodes; carries out a case-based reasoning driven technique to validate/estimate the needs and carries out a utility driven optimal resource allocation plan to minimize resource deficit and deployment time. The CBR driven need validation technique and the integer programming model for optimal resource allocation helps in efficient resource planning. Chapter 4 explains Resource Planning. First the CBR driven need validation technique is presented and then the utility-based integer programming model for optimal resource allocation is illustrated.
- A multilayer encryption and hashing-based scheme is proposed to protect the transmitted resource needs against privacy attacks like—authenticity, eavesdropping and modification attacks. A security aided and group encounter-based PRoPHET routing protocol is also suggested, where security features are incorporated into PRoPHET to provide security against eavesdropping, black hole, identity spoofing and bundle store overflow attacks. Finally, a trust-based Watchdog technique is proposed seamlessly integrated with PRoPHET to provide security against commu-

nication attack like—packet drop and trust/reputation attacks like—bad-mouthing and ballot-stuffing. These security solutions run on the smartphones, carried by volunteers, and contribute towards reliable data dissemination so that resource planning services can be rendered reliably. Chapter 5 presents three security schemes for reliable dissemination of situational data over DTN.
- Chapter 6 winds up the book with conclusion and a discussion on possible works in future direction.

References

1. Luo H, Kravets R, Abdelzaher T (2010) The-day-after networks: a first-response edge-network architecture for disaster relief. Retrieved from http://www.nets-find.net/Funded/DayAfterNet.php. Accessed on Jan 2014
2. Disaster. Oxford living dictionaries. Retrieved from http://www.oxforddictionaries.com/definition/english/disaster. Accessed on July 2017
3. UNISDR Terminology on Disaster Risk Reduction (2009). United Nations international strategy for disaster reduction (UNISDR). Retrieved from http://unisdr.org/files/7817_UNISDRTerminologyEnglish.pdf. Accessed on July 2017
4. The Disaster Management Act (2005). Ministry of Law and Justice, Govt. of India. Retrieved from http://atingl.nic.in/Downloads/THE%20DISASTER%20MANAGEMENT%20ACT%202005.pdf. Accessed on July 2017
5. About disaster management. International Federation of Red Cross and Red Crescent Societies. Retrieved from https://www.ifrc.org/en/what-we-do/disaster-management/about-disaster-management/. Accessed on July 2017
6. Understanding Disasters (2010). National Institute of Disaster Management, Ministry of Home Affairs, Govt. of India. Retrieved from http://nidm.gov.in/PDF/Disaster_about.pdf. Accessed July 2017
7. Disaster Assistance A Guide to Recovery Programs (2005). FEMA. Retrieved from https://www.fema.gov/pdf/rebuild/ltrc/recoveryprograms229.pdf. Accessed on July 2017
8. National Post-Disaster Recovery Planning and Coordination (2016). UNDP. Retrieved from http://www.preventionweb.net/files/53937_undpguidancenotedisasterrecoveryfin.pdf. Accessed on July 2017
9. Disaster Services Start-up Guide for State Service Commissions (2012). Corporation for National and Community Service. Retrieved from https://www.nationalservice.gov/sites/default/files/documents/disasterservices_startupguide.pdf. Accessed on July 2017
10. Post Disaster Needs Assessment (2013) International Recovery Platform. Retrieved from http://www.recoveryplatform.org/pdna/
11. Uddin MYS (2012) Disruption-tolerant networking protocols and services for disaster response communication. Ph.D. dissertation, Graduate College of the University of Illinois, Urbana-Champaign
12. Fall K, Iannaccone G, Kannan J, Silveira F, Taft N (2010) A disruption-tolerant architecture for secure and efficient disaster response communications. In: Proceedings of ISCRAM 2010
13. Paul PS et al (2015) Challenges in designing testbed for evaluating delay-tolerant hybrid networks. In: Proceedings of PerCom workshops 2015. pp 280–283
14. Hossmann T et al (2011) Twitter in disaster mode: security architecture. In: Proceedings of SWID 2011, Article No. 7
15. Mehendale H, Paranjpe A, Vempala S (2011) LifeNet: a flexible ad hoc networking solution for transient environments. SIGCOMM Comput Commun Rev 41(4):446–447

16. Ntareme H, Zennaro M, Pehrson B (2011) Delay tolerant network on smartphones: Applications for communication challenged areas. In: Proceedings of ExtremeCom, 2011, Article No. 14
17. Venkataraman V, Lam S, Acharya HB, Shah H (2011) Delay tolerant networking—a tutorial. Retrieved from www.cs.utexas.edu/~vinodv/files/dtn-tutorial.pdf. Accessed on Jun 2017
18. Fall K, Farell S (2008) DTN: an architectural retrospective. IEEE J Sel Areas Commun 26(5):828–836
19. Warthman F (2012) Delay tolerant networks (DTNs): a tutorial. Retrieved from http://ipnsig.org/wp-content/uploads/2012/07/DTN_Tutorial_v2.04.pdf. Accessed on May 2017
20. Fall K (2003) A delay-tolerant network architecture for challenged internets. In: Proceedings of SIGCOMM 2003, pp 27–34
21. Sun W, Liu C, Wang D (2012) On delay-tolerant networking and its application. Int Proc Comput Sci Inf Technol 51(2012):238–244
22. Misra S, Saha B, Pal S (2016) Delay tolerant routing and applications. Opportunistic mobile networks: advances and applications. Springer, Cham, pp 23–52
23. Massri K (2016) Routing protocols for delay tolerant networks: a reference architecture and a thorough quantitative evaluation. J Sens Actuator Netw 5(2):1–28
24. Keränen A, Ott J, Kärkkäinen T (2009) The ONE simulator for DTN protocol evaluation. In: Proceedings of SIMUTOOLS 2009, Article No. 55
25. Vahdat A, Becker D (2000) Epidemic routing for partially-connected ad hoc networks. Master's thesis, Duke University, Durham, NC, USA
26. Lindgren A, Doria A, Schelén O (2003) Probabilistic routing in intermittently connected networks. SIGMOBILE Mobile Comput Commun Rev 7(3):19–20
27. Spyropoulos T, Psounis K, Raghavendra CS (2005) Spray and wait: an efficient routing scheme for intermittently connected mobile networks. In: Proceedings of SIGCOMM workshop on delay-tolerant networking 2005. pp 252–259
28. Burgess J, Gallagher B, Jensen D, Levine BN (2006) MaxProp: routing for vehicle-based disruption-tolerant networks. In: Proceedings of INFOCOM 2006. pp 1–11
29. Balasubramanian A, Levine B, Venkataramani A (2007) DTN routing as a resource allocation problem. SIGCOMM Comput Commun Rev 37(4):373–384
30. Daly EM, Haahr M. (2007) Social network analysis for routing in disconnected delay-tolerant MANETs. In: Proceedings of MobiHoc 2007. pp 32–40
31. Daly EM, Haahr M. (2007) Social network analysis for routing in disconnected delay-tolerant MANETs. In: Proceedings of MobiHoc 2007. pp 32–40
32. Nelson SC, Bakht M, Kravets R (2009) Encounter-based routing in DTNs. In: Proceedings of INFOCOM 2009. pp 846–854
33. Pentland A, Fletcher R, Hasson A (2004) Daknet: rethinking connectivity in developing nations. Computer 37(1):78–83
34. Seth A et al (2006) Low-cost communication for rural internet kiosks using mechanical backhaul. In: Proceedings of MobiCom 2006. pp 334–345
35. Ginzboorg P et al (2010) DTN communication in a mine. In: Proceedings of ExtremeCom 2010
36. Zhu H (2009) Security in DTNs. Ph.D. thesis, Electrical and Computer Engineering, University of Waterloo. Retrieved from https://uwspace.uwaterloo.ca/bitstream/handle/10012/4348/Haojin%20Zhu's%20Phd%20Thesis.pdf. Accessed on July 2017
37. Ahmad K et al (2011) Classification of internet security attacks. In: Proceedings of INDIACom 2011. pp 1–3
38. Symington S, Farrell S, Weiss H, Lovell P (2011) Bundle security protocol specification. Internet Research Task Force (IRTF). Retrieved from https://www.rfc-editor.org/rfc/rfc6257.txt. Accessed on July 2017
39. Saha S (2015) Design of energy efficient secure delay tolerant networks under QoS bound. Ph.D. thesis, Department of Computer Application, NIT Durgapur, Durgapur, India

Chapter 2
Post-disaster Need Assessment

Post-disaster need assessment deals with the accurate assessment of the need (i.e. demand and utility) for emergency resource at the shelters. While demand signifies the amount of resource required, utility represents the exigency of that requirement. Due to lack of, or imprecise need assessments immediately after a disaster, relief requirements are generally set up based on coarse estimates by logisticians regarding what people would normally need. The effectiveness of this estimation depends on the competencies and experience of the logistician in control, often leading to impromptu allocation of typically scarce emergency resources. Thus, forecasting the exact demand and enumerating the correct utility of emergency resources are inevitable.

In this chapter, first a principal component regression (PCR) model is derived for predicting the exact need for different urgent resources in a shelter, depending on the dynamic influencing situational parameters. Next, a utility function is proposed that quantitatively enumerates the correct utility of each critical resource of different shelters at run-time. Situational parameters of a shelter are uploaded to the shelter-node (any workstation or laptop kept in the shelter) by the emergency supervisor present on site. The shelter-node forecasts the demand for different emergency resources using the proposed PCR model and specifies the utility of such resources using the derived utility function. Such demands and utilities are then communicated to the control station where they are verified and used for making resource allocation plan. The demand forecast model and the utility function assist in accurate *need assessment* for critical resources at various shelters.

2.1 Literature Review

There has been substantial research on predicting the demand for critical resources. In this section, first a few works on emergency resource demand forecasting are reviewed. Next, works on PCR-based demand forecasting are reviewed.

S. Basu et al., *Reliable Post Disaster Services over Smartphone Based DTN*, Smart Innovation, Systems and Technologies 137,
https://doi.org/10.1007/978-981-13-6573-7_2

2.1.1 Emergency Resource Demand Forecasting

Different types of models, namely time series, artificial neural networks and multiple linear regression have been discussed in the literature to predict/model demand for critical resources.

Guo et al. in [1] use the BP neural network algorithm to predict casualties after a disaster, and then gives estimation of the demand for critcal resources depending on knowledge of inventory management. Spencer et al. in [2] demonstrate a multivariate time series approach to modelling and estimating demand in case of emergency. From the perspective of inventory management, Meng [3] also applies BP neural network algorithm to emergency resource inventory prediction problem in coal enterprisers. Sheu [4] suggests a hybrid fuzzy clustering-optimization method for emergency logistics co-distribution in response to the emergency relief demands during the crucial rescue operation. Sun et al. in [5] apply a combination of fuzzy and rough set theories for forecasting emergency resource demands during unconventional emergency events.

All of the above methods predict the demand for critical resources based on past data and do not take into account the several intrinsic and external situational parameters that manipulate such demand. Multiple linear regression (MLR) is a technique that can be used to fit a predictive model to an available data set of a dependent variable (emergency resource demand, in this case) based on a data set of independent variables (situational parameters influencing the demand, in this case). Once developed, such a model can effectively anticipate the value of a dependent variable depending on the present values of the independent variables. In fact, MLR is the most suitable prediction technique for critical resource demands when corresponding data on relevant situational parameters determining the demands are available. There have been a few works on urgent resource demand estimation using MLR. Zhang et al. in [6] establish a model of multiple-linear-regression-case-based reasoning to forecast the demand of various types of emergency resources in the target case. Xiao et al. in [7] propose a forecasting model of unconventional emergence incident's resource demand using MLR.

2.1.2 PCRA-Based Techniques

Regardless of the obvious success of multiple linear regression in different applications, however, the presence of multicollinearity, i.e. high association between the independent variables themselves, results in incorrect forecast of the dependent variable. Principal component regression (PCR) [8, 9] is widely used to deal with multicollinearity among independent variables. PCR is technically a refinement of MLR by using principal component analysis (PCA). PCA transforms the original set of interrelated independent variables to a new set of uncorrelated variables called principal components to avoid the multicollinearity problem and to identify the variables which are the most important for making the forecast. Therefore, it becomes essential to use PCR for predicting the demand for critical resources in this framework. Although there is substantial work done on PCR-based demand prediction for other types of resources, to the best of the knowledge, there is no PCR model for forecasting demands of critical emergency resources. As a result, the literature review is restricted to PCR-based forecasting of non-emergency resources only.

Haque et al. in [10] models water demand by multiple linear regression (MLR) and PCR techniques to guesstimate the future demand for water in the Blue Mountains Water Supply System of New South Wales, Australia. This work satisfies dual objectives: (i) to evaluate the efficiency of principal components as independent variables in multiple linear regression models of forecasting demand of water (i.e. checking the effectiveness of PCR in water demand forecasting); (ii) to assess the performance of developed multiple linear regression and PCR models by estimating and evaluating several goodness of fit statistics. Here, PCs are obtained by undertaking the PCA of ten variables determining water demand. The PCR model is found to outperform the MLR model during the forecasting period.

Saravanan et al. in [11] deal with electricity consumption in India, to forecast future projection of demand for a period of 19 years from 2012 to 2030. The 11 input variables used are Amount of CO2 emission, Population, Per capita GDP, Per capita gross national income, Gross Domestic savings, Industry, Consumer price index, Wholesale price index, Imports, Exports and Per capita power consumption. A new methodology based on artificial neural networks (ANNs) using principal components is used for the forecasting purpose. Historical data of 29 years are used for training and data of the next 10 years used for testing the ANNs. It is shown that ANNs with original data as input variables are more effective than multiple linear regression (based on original data and the principal components).

Ismail et al. in [12] also aim at forecasting the electricity consumption demand in Malayasia. They introduce a new hybrid approach that combines PCR and Back-Propagation Neural Network (BPNN) techniques in order to improve the accuracy of the electricity demand prediction rates. This work includes 13 factors that are related to electricity demand, and data for these factors have been collected in Malaysia. The new combination (PCR–BPNN) solves the problem of collinearity among the input data set, and hence gives reliable results. The work focuses also on the errors that are recorded at the output stage of the electricity prediction models due to changes in the patterns of the input data set. The accuracy and reliability of the results have been improved through the proposed model.

Rajab et al. in [13] and Sousa et al. in [14] propose a multiple-regression-and-principal-component-analysis-based method to improve the prediction accuracy of ozone concentrations. The work in [13] encompasses columnar ozone modelling in the peninsular Malaysia. A method, based on multiple regressions combined with principal component analysis modelling, is used to predict columnar ozone. On the other hand, work in [14] (i) evaluates the relative influence of precursor concentrations and meteorological variables on ozone formation, using principal component analysis; and (ii) predicts next-day hourly ozone concentrations, through a new methodology based on feed forward artificial neural networks using principal components as inputs. Results, in both the works, show that the use of principal components as inputs improved both models' prediction by reducing their complexity and eliminating data collinearity.

Florez et al. in [15] propose a four-step methodology for forecasting disaster impact on the basis of certain influencing variables, namely vulnerability discriminating variables and resilience discriminating variables. This approach uses data analysis techniques such as PCA and multivariate regression analysis. An application case on Peruvian earthquake is proposed to illustrate the benefits of the approach. Such projected disaster impacts can then be used for predicting demand for critical resources.

The above discussion reveals that although there is substantial work done on PCRA-based demand prediction for other kinds of resources, there has been no direct application of PCRA in forecasting the demand for urgent resources based upon the influencing parameters that affects the situations at the shelters. In the light of the above, a PCR model is formulated for estimating post-disaster urgent resource demands in a shelter, based on the influencing situational parameters, which are often inter-correlated.

2.2 Demand Forecast Using Principal Component Regression Model

Predicting the accurate demand for emergency resources in a shelter becomes complicated because parameters influencing these demands continuously change with the changing dynamics of the disaster. This necessitates deriving a precise relationship between the demand for emergency resources and the situational parameters determining the demand, so that it would be possible to forecast the changing demand with the changing nature of the parameters. This section derives a PCR model [16] for predicting the exact need for different critical resources in a shelter, considering the dynamic influence of situational parameters. The resultant model can be used to periodically predict the ever-changing requirements for different urgent resources at shelters with high accuracy. The process of deriving the model involves three different steps—(i) identification of parameters influencing the shelter demands through literature survey, (ii) selection of unrelated parameters from the set of influencing parameters by means of principal component analysis and (iii) development of the forecasting model using MLR on principal components. Figure 2.1 illustrates the step-by-step process of formulating a PCR model for the prediction of urgent resource demands. The resultant model can be used to periodically anticipate the varying demand for different critical resources at different shelters reflecting the changes in the influencing parameters therein.

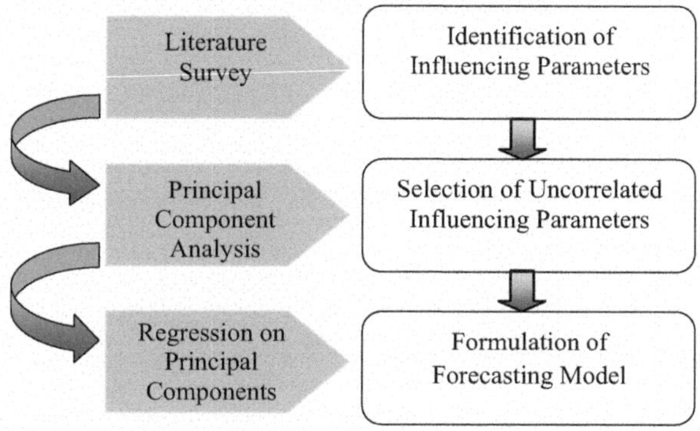

Fig. 2.1 Steps involved in formulating a PCR model

2.2.1 *Identification of Influencing Parameters*

Identification of situational parameters impacting the shelter demands is explained using an example of food resource demand. A comprehensive literature survey is conducted to prepare an accurate and complete list of parameters that determines the demand for food resource in a shelter. Knowledge acquired from the literature survey is further endorsed by the field experts to extract the perfect blend of theory and practice.

According to the "Guidelines for Estimating Food and Nutritional Needs in Emergencies" by WFP/ UNHCR [17], requirement of food resource is decided by the extent of availability of local food and the buying power of the affected population. "Food and Nutrition in Disasters Guidelines" by PAHO/ WHO INCAP [18] mention that prices of food commodities directly influence the purchasing power. "Rapid Joint Needs Assessment Format" by Inter Agency Group [19] and "Emergency Food Security Assessment Handbook" by WFP [20] indicate that the presence of a larger number of adult males in a shelter implies high consumption of food. Infants generally survive on breastfeeding, thus require less external food. These insights give directions for identifying the influencing factors of food resource in a post-disaster scenario.

To accumulate real-life situational information required for post-disaster situational analysis, our project team members visited Nepal after the earthquake in April 2015. They visited several relief shelters at Kathmandu, Sindhupalchowk and Nuwakot. The team interacted with the victims at the shelters, local people, field experts and disaster management volunteers from NGOs like Doctors For You (a team of medical practitioners who provide medical relief to disaster victims, pan India), Qatar Red Crescent and ACF (Action Contre La Faim) International, France. A questionnaire, prepared on the basis of the RAPID Needs Assessment Format [21] provided by the Sphere Project [22], was used primarily to collect the situational information from the responders. The survey was done on six broad categories, viz. Water Supply, Food Supply, Infrastructure, Health and Hygiene, Shelter Conditions, and Economic Activities. A sample questionnaire is provided in Fig. 2.2. Although

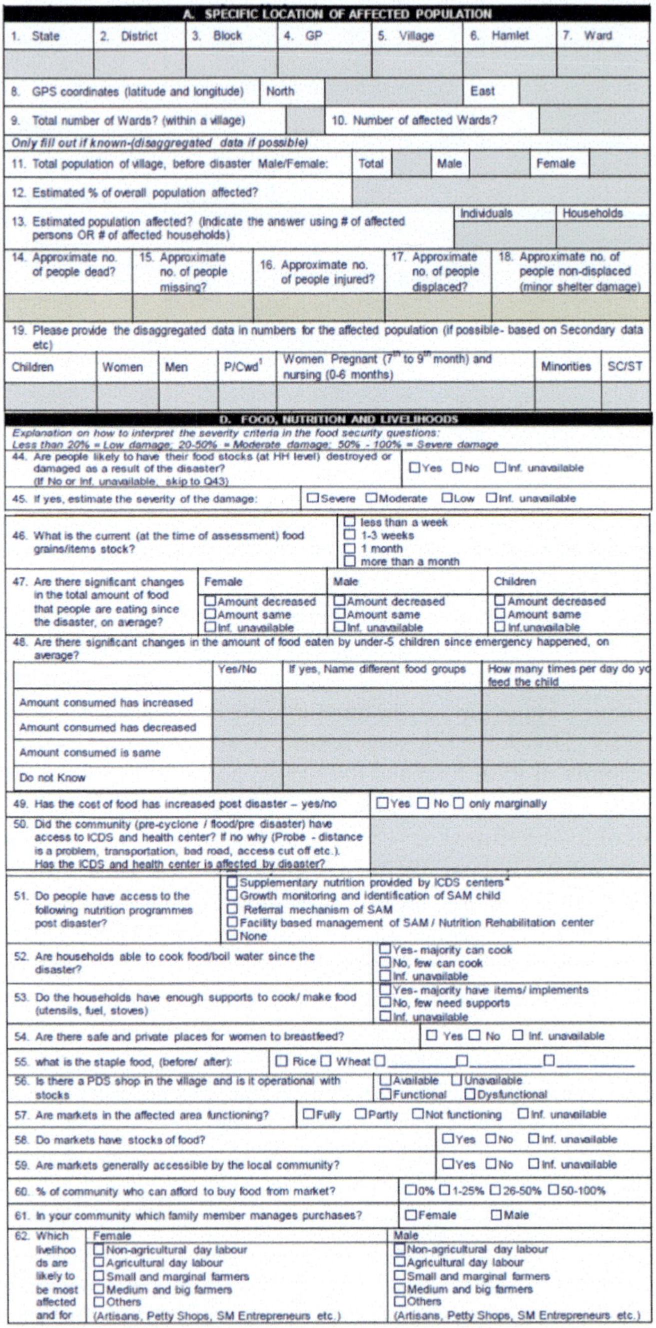

A. SPECIFIC LOCATION OF AFFECTED POPULATION

1. State	2. District	3. Block	4. GP	5. Village	6. Hamlet	7. Ward

8. GPS coordinates (latitude and longitude)	North		East	

9. Total number of Wards? (within a village)		10. Number of affected Wards?	

Only fill out if known-(disaggregated data if possible)

11. Total population of village, before disaster Male/Female:	Total		Male		Female	

12. Estimated % of overall population affected?	

13. Estimated population affected? (Indicate the answer using # of affected persons OR # of affected households)	Individuals	Households

14. Approximate no. of people dead?	15. Approximate no. of people missing?	16. Approximate no. of people injured?	17. Approximate no. of people displaced?	18. Approximate no. of people non-displaced (minor shelter damage)

19. Please provide the disaggregated data in numbers for the affected population (if possible- based on Secondary data etc)

Children	Women	Men	P/Cwd[1]	Women Pregnant (7th to 9th month) and nursing (0-6 months)	Minorities	SC/ST

D. FOOD, NUTRITION AND LIVELIHOODS

Explanation on how to interpret the severity criteria in the food security questions:
Less than 20% = Low damage; 20-50% = Moderate damage; 50% - 100% = Severe damage

44. Are people likely to have their food stocks (at HH level) destroyed or damaged as a result of the disaster? (If No or Inf. unavailable, skip to Q43)	☐Yes ☐No ☐Inf. unavailable
45. If yes, estimate the severity of the damage:	☐Severe ☐Moderate ☐Low ☐Inf. unavailable
46. What is the current (at the time of assessment) food grains/items stock?	☐ less than a week ☐ 1-3 weeks ☐ 1 month ☐ more than a month

47. Are there significant changes in the total amount of food that people are eating since the disaster, on average?	Female	Male	Children
	☐Amount decreased ☐Amount same ☐Inf. unavailable	☐Amount decreased ☐Amount same ☐Inf. unavailable	☐Amount decreased ☐Amount same ☐Inf.unavailable

48. Are there significant changes in the amount of food eaten by under-5 children since emergency happened, on average?

	Yes/No	If yes, Name different food groups	How many times per day do yo feed the child
Amount consumed has increased			
Amount consumed has decreased			
Amount consumed is same			
Do not Know			

49. Has the cost of food has increased post disaster – yes/no	☐Yes ☐ No ☐ only marginally
50. Did the community (pre-cyclone / flood/pre disaster) have access to ICDS and health center? If no why (Probe - distance is a problem, transportation, bad road, access cut off etc.). Has the ICDS and health center is affected by disaster?	
51. Do people have access to the following nutrition programmes post disaster?	☐Supplementary nutrition provided by ICDS centers[2] ☐Growth monitoring and identification of SAM child ☐ Referral mechanism of SAM ☐Facility based management of SAM / Nutrition Rehabilitation center ☐None
52. Are households able to cook food/boil water since the disaster?	☐Yes- majority can cook ☐No, few can cook ☐Inf. unavailable
53. Do the households have enough supports to cook/ make food (utensils, fuel, stoves)	☐Yes- majority have items/ implements ☐No, few need supports ☐Inf. unavailable
54. Are there safe and private places for women to breastfeed?	☐ Yes ☐ No ☐ Inf. unavailable
55. what is the staple food, (before/ after):	☐ Rice ☐ Wheat ☐________ ☐________ ☐________
56. Is there a PDS shop in the village and is it operational with stocks	☐Available ☐Unavailable ☐Functional ☐Dysfunctional
57. Are markets in the affected area functioning?	☐Fully ☐Partly ☐Not functioning ☐Inf. unavailable
58. Do markets have stocks of food?	☐Yes ☐No ☐Inf. unavailable
59. Are markets generally accessible by the local community?	☐Yes ☐No ☐Inf. unavailable
60. % of community who can afford to buy food from market?	☐0% ☐1-25% ☐26-50% ☐50-100%
61. In your community which family member manages purchases?	☐Female ☐Male

62. Which livelihoods are likely to be most affected and for	Female	Male
	☐Non-agricultural day labour ☐Agricultural day labour ☐Small and marginal farmers ☐Medium and big farmers ☐Others (Artisans, Petty Shops, SM Entrepreneurs etc.)	☐Non-agricultural day labour ☐Agricultural day labour ☐Small and marginal farmers ☐Medium and big farmers ☐Others (Artisans, Petty Shops, SM Entrepreneurs etc.)

Fig. 2.2 Snapshots of the used questionnaires [20]

there were some non-responses among the 200 responders interacted with, the average response rate was more than 75%. Some of the selected responders along with their job role and locations are listed in Table 2.1. According to the inputs gathered from the field experts, the raw ration was distributed in different districts of Nepal, on per family basis depending on the number and composition (age–sex distribution) of family members.

Based on literature survey and inputs from field experts, the ten most important parameters (Table 2.2) that control the demand for critical food resource in a shelter are identified. Similar literature survey on other types of emergency resources leads to the identification of influencing parameters for the demand of such resources. A list of situational parameters that control the demand for medical aid requirements is presented in Table 2.3. Since many of these ten influencing parameters, shown in Table 2.2, may exhibit multicollinearity, a systematic PCA approach is required for finding the uncorrelated parameters which account for the maximum variance in the demand data. The objectives of PCA are—

(i) removing correlation among the independent variables and
(ii) reducing the number of parameters to as minimum as possible.

Table 2.1 List of selected responders

Name	Role	Location
Judith Wolff	Program manager, ACF	Trishuali, Nuwakot
Dr Anda Perdan	Coordinator, QRC	Bidur, Nuwakot
Rinku Chakraborty	Coordinator, DFY, QRC	Bidur, Nuwakot
Sunny Borgohain	Coordinator, DFY	Kathmandu
Dr Pritish Jalgayankar	Doctor, DFY	Bidur, Nuwakot
Anjali Dheke	Volunteer, DFY	Kathmandu
Hoshang Khan	Volunteer, DFY	Sindhupalchowk
Mishan Bohora	Volunteer, DFY	Sindhupalchowk
Sajita Gosain	Volunteer, DFY	Kathmandu
Goma Gurung	Volunteer, DFY	Kathmandu
Kedar Adhikary	Villager	Trishuali, Nuwakot
Samir Khan	Villager	Kimthang, Nuwakot
Mr. Lama	Villager	Sindhupalchowk
Mohon Tamang	Villager	Sindhupalchowk
Nikesh Adhikary	Villager	Thamel, Kathmandu
Lakib Thakuri	Villager	Trishuali, Nuwakot
Ganesh Gurung	Villager	Trishuali, Nuwakot
Amrit Lama	Villager	Kimthang, Nuwakot
Marina Saba	Villager	Bidur, Nuwakot

Table 2.2 Situational parameters influencing demand for food resources

Parameter Id	Description
ISP_1	Average number of households residing in the shelter
ISP_2	Average household size
ISP_3	Percentage of adult males
ISP_4	Percentage of adult females
ISP_5	Percentage of infants
ISP_6	Percentage of households below national poverty line
ISP_7	Percentage of families who can buy food
ISP_8	Average no. of local shops functioning
ISP_9	Percentage increase in food price
ISP_{10}	Percentage of total demand that can be met from local shops

Table 2.3 Situational parameters influencing demand for medical resources

Parameter Id	Description
ISP_1	Nature of disaster (floods, earthquake, cyclone, conflict, etc.)
ISP_2	Extent of disaster (area under water-floods, magnitude of earthquake, etc.)
ISP_3	Number of persons affected
ISP_4	Number of persons injured
ISP_5	Number of persons displaced
ISP_6	Number of health facilities affected and non-functional
ISP_7	Existing development deficiency
ISP_8	Number of pregnant mothers
ISP_9	Number of persons under-five
ISP_{10}	Number of persons in the reproductive age group
ISP_{11}	Number of persons having major disease condition (Malnutrition, Tuberculosis, Hypertension, Diabetes, etc.)
ISP_{12}	Number of persons having endemic diseases (Malaria, Measles, Diarrhoea, etc.)

2.2.2 Selection of Uncorrelated Influencing Parameters

The set of situational parameters that control the demand (independent variables) for an emergency resource in a shelter, few of which may be correlated, is represented as $\boldsymbol{ISP} = (\boldsymbol{ISP_1}, \boldsymbol{ISP_2}, \ldots, \boldsymbol{ISP_p})$. The data set on these parameters for $\boldsymbol{n}$ time points (time points may be days, weeks or months) is represented as $\{ISP_{t_i1}, ISP_{t_i2}, \ldots, ISP_{t_ip}\}_{i=1}^{n}$. PCA is applied to find a linear transformation of $\boldsymbol{ISP}$s into a set of principal components ($\boldsymbol{PC}$s), denoted by $(\boldsymbol{PC}_1, \boldsymbol{PC}_2, \ldots, \boldsymbol{PC_p})$. The $\boldsymbol{PC}$s can be obtained as linear combination of the $\boldsymbol{ISP}$s, as shown below:

$$\boldsymbol{PC_1} = c_{11}\boldsymbol{ISP_1} + c_{12}\boldsymbol{ISP_2} + \cdots + c_{1n}\boldsymbol{ISP_p} = \sum_{j=1}^{n} c_{1j}\boldsymbol{ISP_j}$$

$$\boldsymbol{PC_2} = c_{21}\boldsymbol{ISP_1} + c_{22}\boldsymbol{ISP_2} + \cdots + c_{2n}\boldsymbol{ISP_p} = \sum_{j=1}^{n} c_{2j}\boldsymbol{ISP_j}$$

$$\vdots$$

$$\boldsymbol{PC_p} = c_{p1}\boldsymbol{ISP_1} + c_{p2}\boldsymbol{ISP_2} + \cdots + c_{pn}\boldsymbol{ISP_p} = \sum_{j=1}^{n} c_{pj}\boldsymbol{ISP_j}$$

where the coefficients C_{ij}, for i =1, 2, ..., p and j =1, 2, ..., n, are components of the eigenvectors $(\boldsymbol{e}_1, \boldsymbol{e}_2, \ldots, \boldsymbol{e_p})$, derived from correlation matrix of the ***ISP***s. The ***PC*** set maintains the properties—(i) elements are uncorrelated with each other; and (ii) $\boldsymbol{PC_1}$ to $\boldsymbol{PC_p}$ are arranged in descending order in terms of variance of the original ***ISP***s.

Eigenvalue λ of the correlation matrix $\boldsymbol{C}$ is calculated using the equation $|\boldsymbol{C} - \lambda\boldsymbol{I}| = 0$, where $\boldsymbol{I}$ is the identity matrix. The above equation results in p eigenvalues, corresponding to the p ***PC***s. These eigenvalues are the actually variances of the ***PC***s. Eigenvectors, e_i, $i = 1, 2, \ldots, p$, corresponding to each eigenvalue are calculated using the equation $|\boldsymbol{C} - \lambda\boldsymbol{I}|\boldsymbol{e_i} = 0$. From the properties of ***PC***s mentioned earlier, the first principal component, $\boldsymbol{PC}_1$, has maximum variance and therefore accounts for as much variation in the data set as possible. The last principal component **PC*p*** has the least variance. Therefore, the first few ***PC***s, which are essentially uncorrelated and account for the maximum variance (say more than 80% of the variance in the original data), can be used to represent the available data set on the influencing situational parameter values. The ***ISP***s with nonzero c_{ij} in the selected ***PC***s comprises the set of uncorrelated ***ISP***s say, $(\boldsymbol{ISP}'_1, \boldsymbol{ISP}'_2, \ldots, \boldsymbol{ISP}'_{\boldsymbol{m}})$ with $m < p$. Thus, a lesser number of ***ISP***s are to be dealt with now. Figure 2.3 illustrates the concept on the original influencing parameters $(\boldsymbol{ISP}_1, \boldsymbol{ISP}_2, \ldots, \boldsymbol{ISP_p})$ PCA results in two principal components, $\boldsymbol{PC_1}$ and $\boldsymbol{PC_2}$ that are responsible for the maximum variance (say more than 80%) in the data and finally find the uncorrelated influencing parameters $(\boldsymbol{ISP}_1, \boldsymbol{ISP}_2, \boldsymbol{ISP}_4, \boldsymbol{ISP}_5, \boldsymbol{ISP}_7)$.

Since the current status of the ***ISP***s varies over time, i.e. the data set $\left\{ISP_{t_i1}, ISP_{t_i2}, \ldots, ISP_{t_ip}\right\}_{i=1}^{n}$ keeps on changing, the relative importance of the ***ISP***s may not remain same forever. Therefore, PCA needs to be repeated periodically with the most recent data set $\left\{ISP_{t_i1}, ISP_{t_i2}, \ldots, ISP_{t_ip}\right\}_{i=1}^{n}$. The interval of such repetition depends on the rate at which the values of the influencing situational parameters change. Thus, PCA helps in removing multicollinearity and reducing the dimension of the available data set. The subsequent PCR technique leads to the efficient prediction of emergency resource demands through proper choice of the uncorrelated influencing parameters.

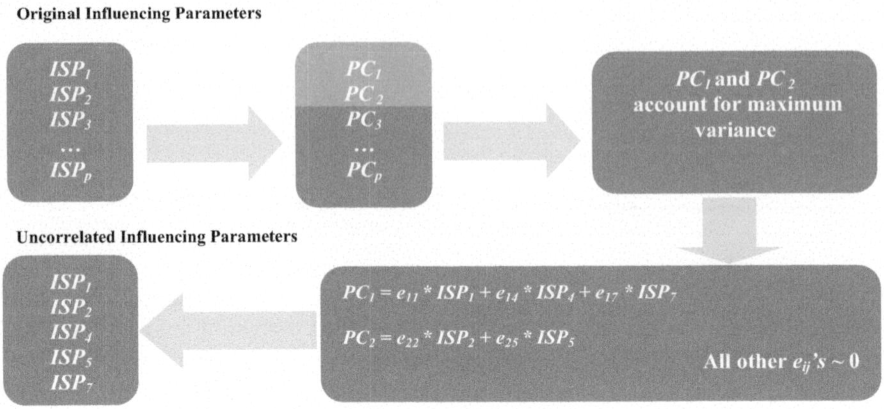

Fig. 2.3 Selection of uncorrelated influencing parameters using PCA

2.2.3 *Formulation of Forecasting Model*

After choosing the uncorrelated ***ISP***s, a forecast model for estimating demand $\boldsymbol{D_{rs}}$ for the rth emergency resource in the sth shelter is formulated. MLR is used on the selected ***PC***s to derive a predictive model to an available data set $\left\{D_{t_i rs}, ISP'_{t_i 1}, ISP'_{t_i 2}, \ldots, ISP'_{t_i m}\right\}_{i=1}^{n}$ on the $\boldsymbol{D_{rs}}$ and $(\boldsymbol{ISP'_1}, \boldsymbol{ISP'_2}, \ldots, \boldsymbol{ISP'_m})$ for $\boldsymbol{n}$ time points. The multiple linear regression model takes the form as follows:

$$D_{t_i rs} = \beta_1 ISP'_{t_i 1} + \beta_2 ISP'_{t_i 2} + \cdots + \beta_m ISP'_{t_i m} + \varepsilon_{t_i}, i = 1, 2, \ldots, n$$
$$\text{i.e., } \boldsymbol{D_{rs}} = \boldsymbol{ISP'\beta} + \boldsymbol{\varepsilon}$$

where $\boldsymbol{D_{rs}} = \left(D_{t_1 rs}, D_{t_2 rs}, \ldots, D_{t_n rs}\right), \boldsymbol{ISP'} = \left(\boldsymbol{ISP'_1}, \boldsymbol{ISP'_2}, \ldots, \boldsymbol{ISP'_m}\right), \boldsymbol{\beta} = (\beta_1, \ldots, \beta_m)^T$ and $\boldsymbol{\varepsilon} = \left(\varepsilon_{t_1}, \ldots, \varepsilon_{t_n}\right)$. Here, $\boldsymbol{\beta}$ and $\boldsymbol{\varepsilon}$ are the vectors of regression coefficients and errors, respectively. The regression coefficients are estimated using least squares method by minimizing the sum of squares of errors $\sum_{i=1}^{n} \varepsilon_{t_i}^2$ which is expressed as $\sum_{i=1}^{n} \varepsilon_{t_i}^2 = \sum_{i=1}^{tn}\left(D_{t_i rs} - \left(\beta_1 ISP'_{t_i 1} + \beta_2 ISP'_{t_i 2} + \ldots + \beta_m ISP'_{t_i m}\right)\right)^2$. Solving the normal equations gives $\widehat{\beta}_1, \widehat{\beta}_2, \ldots \widehat{\beta}_m$ as the estimates of $\widehat{\beta}_1, \widehat{\beta}_2, \ldots \widehat{\beta}_m$. The fitted PCR model for forecasting demand of an emergency resource in a shelter can be expressed as

$$\widehat{D}_{t_i rs} = \hat{\beta}_1 ISP'_{t_i 1} + \hat{\beta}_2 ISP'_{t_i 2} + \cdots + \hat{\beta}_m ISP'_{t_i m} + \varepsilon_i, \quad i = 1, 2, \ldots, n$$
$$\text{i.e., } \widehat{D}_{\boldsymbol{rs}} = \boldsymbol{ISP\hat{\beta}} + \varepsilon \tag{2.1}$$

In the above PCR model, the ***ISP***s are uncorrelated and have reduced dimension thus enabling the model to predict $\boldsymbol{D_{rs}}$ with significant accuracy. If values of the uncorrelated influencing parameters for future time points (t_{n+1} onwards) are available, the fitted model can be used to make a prediction of the demand for an emergency resource in that shelter for those future time points.

2.2.4 Performance Analysis

The efficiency of the proposed PCR-based demand forecasting model is evaluated through quantitative analysis in this section.

2.2.4.1 Simulation Environment

A major earthquake of magnitude—7.8 struck central Nepal on April 25, 2015, leaving calamitous impacts across the country. The most affected districts were Sindhupalchowk, Kathmandu, Nuwakot, Bhaktapur, Dhading, Rasuwa and Gorkha, as shown in Fig. 2.4a [23]. The Google Map of water, food, shelter and medical resources for Nepal earthquake [24] marks the shelters and medical relief centres set in Kathmandu and its neighbouring districts like Nuwakot, Sindhupalchowk, Bhakatpur and others, shown in Fig. 2.4b.

Out of the worst hit districts mentioned above, Kathmandu, Nuwakot and Sindhupalchowk are considered as the study area to set up the simulation environment based on the above snapshot.

2.2.4.2 Simulation Data

Resources were distributed in the affected districts of Nepal from different government and non-government organizations. Such aids had been provided for quite a few weeks following the disaster. Data on food resource distribution were collected from the Nepal Disaster Risk Reduction Portal provided by Government of Nepal [25] and the Nepal Earthquake Recovery Monitoring Assessment Report provided by Shelter Cluster, Nepal in association with International Federation of Red Cross and Red Crescent Societies [26]. Resource distribution data are derived from [24, 25] and [26] for three districts, namely Kathmandu, Sindhupalchowk and Nuwakot for three specific resources—Rice (R_1), Sugar (R_2) and Biscuits (R_3). Table 2.4 presents the average amount of resources allocated per shelter in Kathmandu. Similar tables are prepared for shelters in Sindhupalchowk and Nuwakot. The average amount of an allocated resource in a shelter at a particular time point is assumed to be the minimum requirement at that time point. This minimum requirement is considered as the observed demand.

The number and composition (age–sex–health condition, etc.) of victims taking temporary refuge in the shelters keep on changing over time due to events like closing of other nearby shelters. For example, several shelters were closed due to the May 12 aftershock in Nepal. As a result, the demographic and economic status of the

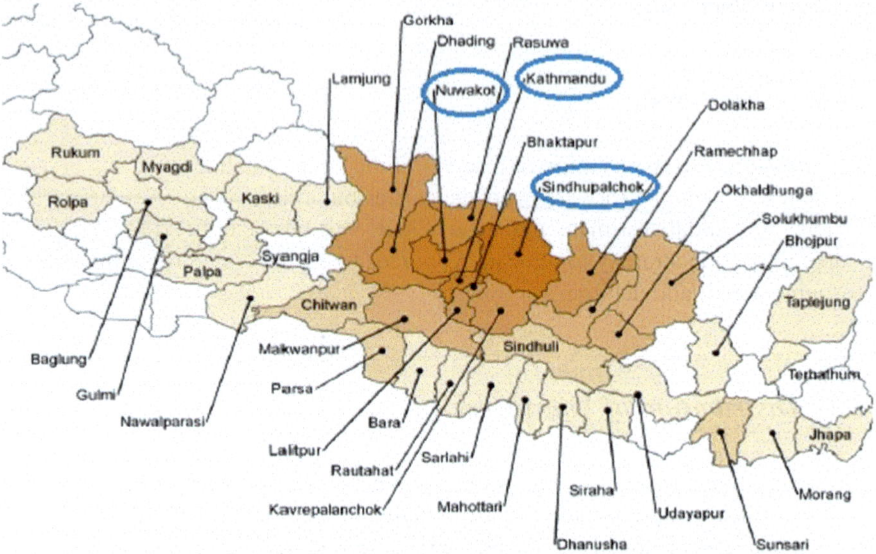

(a): Devastation Map showing the worst hit districts [23]

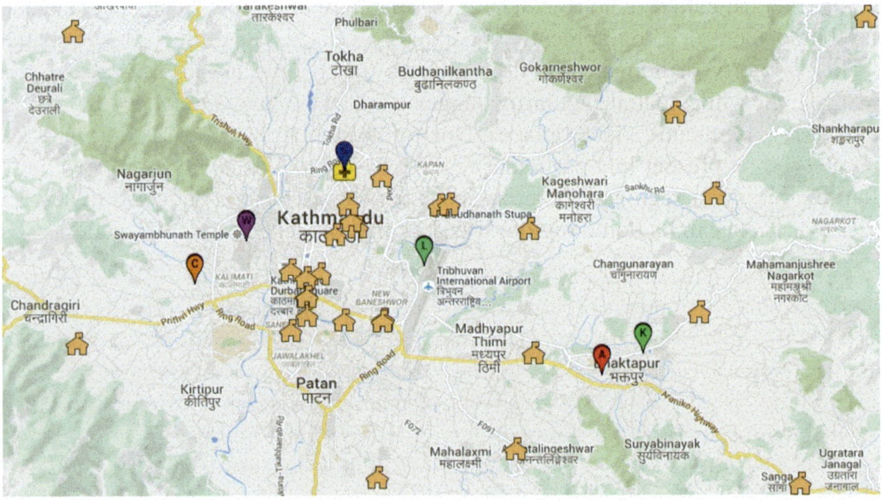

(b): Google Map showing shelters and relief centres [24]

Fig. 2.4 Relevant maps of Nepal earthquake

Table 2.4 Observed Demand Per Shelter In Kathmandu

Week	1	2	3	4	5	6	7	8	9	10	11	12	13	14
R_1(quintals)	18	21	28	642	474	47	63	49	32	36	603	390	192	56
R_2(quintals)	2	2	3	10	9	8	8	6	2	2	10	5	1	1
R_3(no. of cartons)	109	121	218	1000	496	93	11	70	83	76	215	175	150	29

inhabitants of the shelters changed dynamically and hence values of the parameters impacting demand for resources in the shelters also changed over time. As mentioned before, project team members visited several shelters set up in the disaster-affected districts like Kathmandu, Sindhupalchowk and Nuwakot. After extensive interactions with various responders like victims, shelter volunteers, disaster managers of different disaster management agencies and representatives of NGOs, they were able to gather information about the different situational parameters that influence demand for critical resources in the shelters. Values of the ten influencing situational parameters $ISP_1 - ISP_{10}$ (Table 2.2) that impact the demand for emergency food resources in a shelter are derived from the gathered information. Table 2.5 lists the average values of such of ISPs, collected from all the available sources, in shelters of Kathmandu for the period 3 May–31 May 2015. Similar tables are generated for shelters in Nuwakot and Sindhupalchowk. The data sets presented in Tables 2.4 and 2.5 are used as historical data for the PCR-based demand forecasting.

Table 2.5 Average values of different influencing parameters at shelters in Kathmandu

Week	Parameters									
	ISP_1	ISP_2	ISP_3	ISP_4	ISP_5	ISP_6	ISP_7	ISP_8	ISP_9	ISP_{10}
1	2046	5	49	51	5	10	25	8	22	9
2	2015	7	56	44	6	13	33	15	8	27
3	1978	6	45	55	6	11	45	12	15	14
4	2505	6	50	50	5	9	52	20	2	32
5	2768	7	46	54	5	5	59	18	7	30
6	2004	7	57	43	6	10	46	9	18	11
7	1808	7	48	52	6	10	45	9	18	14
8	1983	7	52	48	5	12	40	18	7	29
9	2564	5	40	60	8	12	34	19	6	30
10	2278	7	57	43	8	14	30	7	25	9
11	2065	7	42	58	7	6	55	12	15	20
12	2524	6	59	41	8	4	60	5	30	7
13	2256	7	56	44	5	7	52	20	5	32
14	1880	7	49	51	8	15	29	13	10	22

2.2.4.3 Simulation Set-up

For the PCR-based demand prediction, it is assumed that the average resource demands and the average values of the influencing parameters, presented in Tables 2.4 and 2.5, respectively, are applicable to all the shelters in Kathmandu. XLSTAT [27] is used to perform PCA on the data set in Table 2.5 for identifying the uncorrelated ***ISP***s that accounts for maximum variability in the demand data set. Once identified, data on these ***ISP***s and data on resource demands are fed in XLSTAT to derive a PCR model for estimatingresource demands in a representative shelter of Kathmandu. Accuracy of the model is evaluated with several performance evaluation metrics, discussed in the next section, using demand data for the first 14 weeks. Similar models are derived for shelters of Nuwakot and Sindhupalchowk. The derived models are used to estimate demand for resources, namely R_1, R_2, and R_3 at week 15 for the shelters in each region.

2.2.4.4 Simulation Metrics

Three simulation metrics, as in [10], are used for assessing the performance of the PCR-based demand forecasting model. The metrics are defined as below:

$$\boldsymbol{Percentage\ Bias}(\boldsymbol{PB}) = \frac{1}{\sum_{i=1}^{n} D_{t_i rs}} \sum_{i=1}^{n} \left(D_{t_i rs} - \widehat{D}_{t_i rs} \right) \times 100$$

where $D_{t_i rs}$ and $\widehat{D}_{t_i rs}$ denote observed demand and PCR forecast demands, respectively. The ideal value of PB is 0 which indicates a perfect agreement between the forecast and observed demands. Negative and positive values indicate over- and underestimation, respectively.

$$\boldsymbol{Nash{-}Sutcliffe\ Efficiency}(\boldsymbol{NSE}) = 1 - \left[\frac{\sum_{i=1}^{n} \left(\boldsymbol{D}_{t_i rs} - \widehat{\boldsymbol{D}}_{t_i rs} \right)^2}{\sum_{i=1}^{n} \left(\boldsymbol{D}_{t_i rs} - \overline{\boldsymbol{D}_{rs}} \right)^2} \right]$$

where $\overline{D_{rs}}$ denotes the mean of observed demands. The ideal value is 1 indicating a perfect match of forecast to observed demands; a value between 0 and 1 signifies an acceptable performance.

$$\boldsymbol{Accuracy\ Factor}(\boldsymbol{AF}) = 10 \times \frac{1}{\boldsymbol{n}} \sum_{i=1}^{n} \left| \log \frac{\widehat{\boldsymbol{D}}_{t_i rs}}{\boldsymbol{D}_{t_i rs}} \right|$$

The ideal value is 1, indicating perfect agreement between the estimated and observed values. The smaller the value, the more accurate are the estimation.

2.2.4.5 Simulation Results

In this first step of simulation, the principal components (PCs) are identified through PCA, followed by formulation of a PCR model based on the identified PCs. Subsequently, the effectiveness and suitability of the PCR model in forecasting future demands is evaluated. Finally, the derived PCR model is used to predict the demands of resources for a future time point.

- ***Steps in PCA***: Table 2.4 is used to calculate the correlation between different influencing parameters ($ISP_1 - ISP_{10}$). In Table 2.6, the correlation values that are greater than 0.5 and less than 0.5 are considered to be high (highlighted in bold). These observations imply the existence of multicollinearity among the influencing parameters that determine the demand for emergency food resources in a shelter. *PCA* is performed on the *ISP*s to find out the PCs that are uncorrelated and are responsible for the maximum variance in the observed data. The results of the PCA are illustrated in Fig. 2.5a where eigenvalues and cumulative variances are plotted against the PCs. The figure demonstrates that the eigenvalues and the variances of the first five PCs are significant which explain 99% of the total variance of variables. In fact, PC_1 and PC_2 are dominant in comparison with the other PCs and are selected for factorial analysis to determine the contribution of an *ISP* within a selected PC. Higher correlation of an *ISP* with the PC indicates greater contribution. The results of the factorial analysis are presented in Fig. 2.5b, where *ISP*s of PC_2 are plotted against PC_1. The figure shows that the contribution of $ISP_1, ISP_2, ISP_3, ISP_6$ and ISP_8 is the higher compared to the other *ISP*s. The contribution of these parameters in PC_1 and PC_2 explains more than 80% of variance. These *ISP*s are used for deriving the principal component regression model.
- ***Deriving the PCR Model***: Next, the multiple linear regression equation is established based on ISP_1, ISP_2, ISP_3, ISP_6 and ISP_8. Tables 2.3 and 2.4 and Eq. (2.1) are used to obtain the regression equation for forecasting the average demand for the resource R_1 in shelters of Kathmandu. The regression equation (PCR model), derived using least squares method as illustrated in Sect. 2.2.3, is as follows:

$$\widehat{D}_{R_1s} = -1152.80 + 1.087ISP_1 + 46.052ISP_2 - 26.47ISP_3 + 3.30ISP_6 - 0.78ISP_8 \tag{2.2}$$

where $\widehat{D}_{R_1s}$ is the estimated (forecast) demand for resource R_1 in shelter S. Next, the standardized regression coefficients are obtained with respect to Eq. (2.2), to indicate the influence of each *ISP* on the PCR forecast demand. The standardized regression coefficients are plotted against the *ISP*s in Fig. 2.6. It is observed that the influence of ISP_1 and ISP_2 on the forecast demand is significant. Variations in these influencing parameter values affect the demand to a greater extent. PCR models similar to Eq. (2.2) are derived for all resources at shelters of Kathmandu, Nuwakot and Sindhupalchowk. The derived models are used to compute the demand for each resource at shelters of each region for the first 14 weeks.

Table 2.6 Correlation matrix of influencing parameters

Influencing parameters	ISP_1	ISP_2	ISP_3	ISP_4	ISP_5	ISP_6	ISP_7	ISP_8	ISP_9	ISP_{10}
ISP_1	**1**	−0.306	−0.067	0.067	0.048	0.171	**−0.601**	−0.371	0.237	**−0.920**
ISP_2		**1**	0.356	−0.356	−0.080	−0.302	0.299	0.324	−0.266	0.329
ISP_3			**1**	**−1.000**	−0.028	0.046	−0.027	0.064	−0.150	0.140
ISP_4				**1**	0.028	−0.046	0.027	−0.064	0.150	−0.140
ISP_5					**1**	0.065	−0.111	−0.116	0.111	−0.189
ISP_6						**1**	**−0.976**	−0.402	0.281	**−0.540**
ISP_7							**1**	0.476	−0.372	**0.508**
ISP_8								**1**	**−0.968**	**0.890**
ISP_9									**1**	**−0.695**
ISP_{10}										**1**

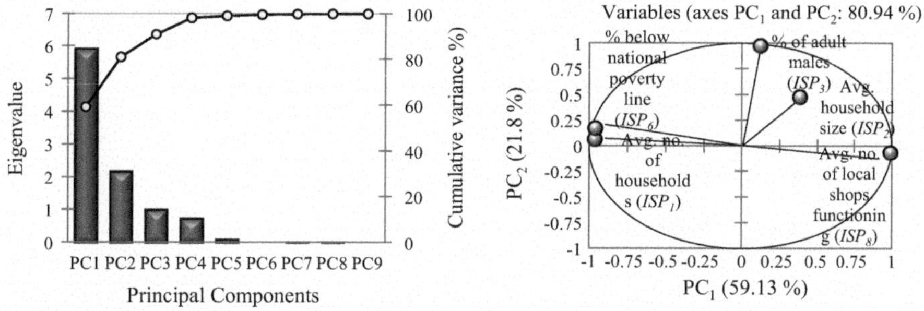

(a) Eigenvalues and variances explained by PCs

(b) Factorial plan of PCA

Fig. 2.5 Steps involved in PCA

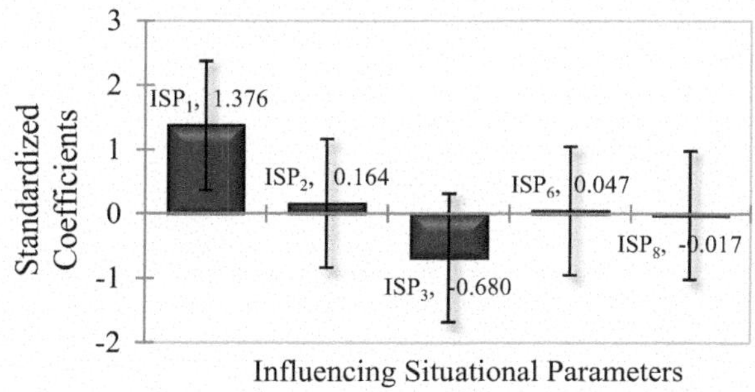

Fig. 2.6 Standardized coefficients (95% confidence interval)

Evaluation of the PCR Model: Performance of the derived PCR models with respect to the observed demands of resource R_1 is plotted in Fig. 2.7. Figure 2.7a shows that the average *PB* is 2.75, average *NSE* is 0.26 and average AF is 1.99, all of which are within the tolerable limits as indicated in Sect. 2.2.4.4. It is further observed from Fig. 2.7b that the average relative error is 2.97% indicating the PCR forecast demands to be sufficiently close to the actual demands. The above results reflect that the derived PCR model can successfully estimate and forecast future demands.

– ***Forecasting Demands for Week* 15**: Next, the derived PCR models are used to estimate the demand for resources, R_1, R_2, and R_3 at shelters of Kathmandu, Nuwakot and Sindhupalchowk for week 15, as described in Sect. 2.2.3. Such forecasts are listed in Table 2.7. These forecast demands are then communicated to the control station using the opportunistic knowledge sharing scheme (discussed in the next chapter), where they are authenticated and used for resource planning.

In this work, a PCR model is designed with an aim to forecast post-disaster resource demands based on dynamically changing situational parameters. Experimental results justify that the model projects resource demands in the shelters with noticeable accuracy (accuracy factor nearly 1). The derived model is used to predict the future demand for emergency resources at different shelters which are then com-

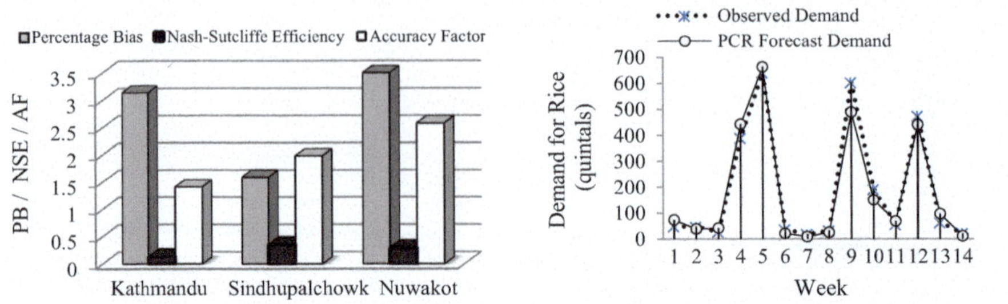

(a) PB, NSE, and AF of PCR forecast demand (b) PCR forecast demand vs. observed demand for first 14 weeks in shelters of Kathmandu

Fig. 2.7 Performance of PCR model for resource R_1

Table 2.7 PCR forecast resource demands for week 15

Resource	Demand in shelters		
	Kathmandu	Sindhupalchowk	Nuwakot
R_1	52	72	66
R_2	1	4	3
R_3	30	51	49

municated to the control station and are used for resource planning after sufficient validation check.

While responding to a post-disaster relief operation with limited availability of resources, identifying the utilities of those resources at different shelters is of immense importance for effective allocation strategies. While demand represents the quantity of resource required, utility describes the urgency of that requirement. However, quantifying a subjective notion like utility is fairly challenging. In the next section, we derive a utility function that dynamically and quantitatively specifies the utility of each emergency resource and use this function to measure the exigency of such resources at each individual shelter.

2.3 Utility Enumeration Through Utility Function

Enumerating the correct utility of such emergency resources is challenging in a dynamic post-disaster scenario, where demands are constantly changing and supplies drop in at an uncertain rate. Therefore, a utility function that quantitatively enumerates the utility of each emergency resource in a dynamic manner needs to be derived and used to measure the exigency of such resources at each individual shelter. This section derives a utility function [28] that dynamically quantifies the correct utility of each emergency resource at different shelters.

According to the notion of utility defined in classical Economics [29, 30], utility of a critical resource is identified as its importance with which the demand in a shelter at a point of time is to be treated. Utility is different from demand in the sense that demand for a resource specifies the quantity needed, whereas utility of the resource is the exigency with which it is needed. For example, a shelter having demand for tents that has not been supplied with any tent for the past 36 h has definitely higher utility of tents than a shelter where demand for the tents has been periodically met. Hence, continuous enumeration of utilities of different resources is crucial to depict the time-varying importance of various resource needs in a shelter.

2.3.1 Deriving the Utility Function

Utility of a resource is expressed as a function of the number of units of that resource being consumed. It is represented as a mathematical expression that enumerates the utility gained by consuming different units of the commodity. Therefore, utility u can be defined as a function of consumption, x. The utility function is represented as

$$u = f(x)$$

Table 2.8 illustrates the utility of clams based on the consumption of different units of clams over a given time period [31]. It is observed from the table that utility rises till the sixth unit of consumption and reduces beyond this point. As a result, the corresponding utility curve (Fig. 2.8) is concave in nature.

To derive a general mathematical expression for utility functions, standard curve fitting technique is employed on sample conventional utility data sets derived from [31–38] (e.g. no. of movies/week, no. of roller coaster rides/day, transmission rate/hour, no. of chocolate bars eaten/ hour, etc.). Curve fitting reaches the conjecture that the following model fits best as a utility function

Table 2.8 Utility obtained by consuming different units of clams

Units Consumed	0	1	2	3	4	5	6	7	8	9
Utility	0	15	28	39	48	55	60	60	59	56

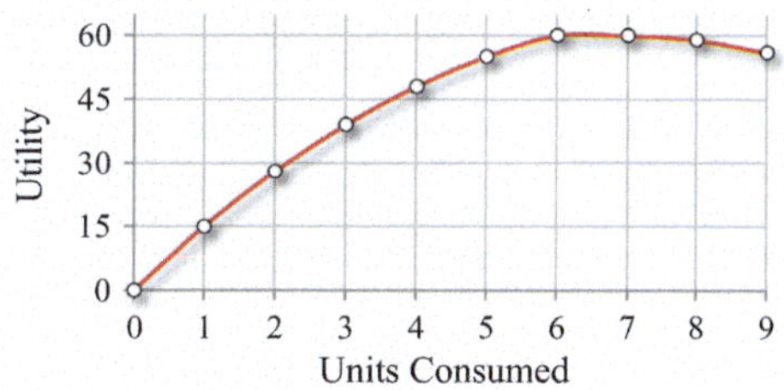

Fig. 2.8 Utility of clams against consumption

$$u = ax^3 + bx^2 + cx + d \tag{2.3}$$

The fitted model gives predicted utility values sufficiently close to the actual values. Three goodness of fit statistics, namely R-square, Adjusted R-square and RMSE are used to evaluate the fitness of the model. Significance of the above-mentioned statistics is described below:

- ***R-square***: It is the square of the correlation between actual values and forecast values. A value closer to 1 indicates a better fit.
- ***Adjusted R-square***: It is the R-square statistic adjusted on the basis of residual degrees of freedom. This is the best indicator of the fit. A value closer to 1 indicates a better fit.
- ***RMSE (Root Mean Squared Error)***: It is the estimate of the standard deviation of the random component in the data. A value closer to 0 indicates a better fit.

Table 2.9 summarizes values of these goodness of fit statistics for each of the above-mentioned utility data sets from [31–38] for the utility function in (2.3). The table shows that the *R-square* as well as

Adjusted R-square values are close to 1 and *RMSE* values are close to 0 for each data set. The above utility function (2.3) is applied on a number of standard utility data sets and similar results as in Table 2.9 have been observed. This is quite obvious owing to the similar nature of all utility data sets. Plotting the utility values available from other utility data sets against the corresponding utility curves drawn on these data sets using (2.3) shows sufficiently small deviations. Thus, Eq. (2.3) where utility is expressed as a function of consumption can be considered as a general utility function.

In the above discussion, utility u_{rs} of an emergency resource, say, r, in a shelter, say, s, is considered as a function of total consumption (or, allocation), say, x_{rs}, of that resource (r) in the shelter (s). However, in post-disaster relief operations, resources are allocated to shelters at irregular intervals (may be non-uniform). This implies that allocation of a resource, x_{rs}, in a shelter, s, varies with time, i.e. x_{rs} is a function of time t. Therefore, it is evident that u_{rs} depends on x_{rs} which in turn depends on allocation time t, because x_{rs} can vary only if t varies. Hence,

Table 2.9 Goodness of fit statistics for the model $u = ax^3 + bx^2 + cx + d$

Dataset	R-Square	Adjusted R-square	RMSE
1	1	1	0.0289
2	0.9995	0.9993	0.5725
3	0.9929	0.9714	0.6715
4	0.9994	0.9992	0.1212
5	1	1	2.521 e−4
6	0.9991	0.9988	0.2188
7	1	1	1.239 e−4
8	0.9905	0.9809	0.6225

$$u = f(x) = f(g(t))$$

Therefore, the next task would be to examine the nature of the relation between x_{rs} and t. Since resources arrive at shelters at non-uniform time points, $t_1, t_2, \ldots, t_n$, it is reasonable to use Lagrange's Interpolation formula to find the dependence of x_{rs} on t. Hence, x_{rs} can be represented as

$$x_{rs} = \frac{(t-t_2)(t-t_3)\ldots(t-t_n)}{(t_1-t_2)(t_1-t_3)\ldots(t_1-t_n)}x_{1rs} + \frac{(t-t_1)(t-t_3)\ldots(t-t_n)}{(t_2-t_1)(t_2-t_3)\ldots(t_2-t_n)}x_{2rs}$$
$$+ \cdots + \frac{(t-t_1)(t-t_2)\ldots(t-t_{n-1})}{(t_n-t_2)(t_n-t_2)\ldots(t_n-t_{n-1})}x_{nrs}$$
$$\text{i.e., } x_{rs} = \sum_{i=1}^{n}\prod_{\substack{j=1\\ j\neq 1}}^{n}\frac{t-t_j}{t_i-t_j}x_{irs} \tag{2.4}$$

From Eqs. (2.3) and (2.4),

$$u_{rs} = a\left(\sum_{i=1}^{n}\prod_{\substack{j=1\\ j\neq 1}}^{n}\frac{t-t_j}{t_i-t_j}x_{irs}\right)^3 + b\left(\sum_{i=1}^{n}\prod_{\substack{j=1\\ j\neq 1}}^{n}\frac{t-t_j}{t_i-t_j}x_{irs}\right)^2$$
$$+ c\left(\sum_{i=1}^{n}\prod_{\substack{j=1\\ j\neq 1}}^{n}\frac{t-t_j}{t_i-t_j}x_{irs}\right) + d \tag{2.5}$$

Therefore, utility of an emergency resource is its urgency at a particular time point. Urgency of a resource in a shelter decreases with time if that specific resource is periodically allocated to that shelter; urgency increases if that resource is not allocated after a certain time, leading to starvation for that resource.

2.4 Enumerating Resource Utility—An Illustrative Example

This section illustrates the utility enumeration process for an emergency resource in a shelter using the utility function (2.5). An example scenario is assumed where tents are continuously allocated to a shelter for a period of time and the utility of tents on different days in that shelter are enumerated. The corresponding example data set for days (t) and cumulative number of allocated tents (x_{rs}) on days (t) are provided in Table 2.10. Now putting these values in Eq. (2.4), x_{rs} values for different days are obtained. Putting these x_{rs} values in (2.5), with $a = 0.1086$, $b = -4.122$, $c = 39.97$,

Table 2.10 Cumulative number of allocated tents & utility of tents on different days

Day (t)	Continuous allocation		Allocation stopped after day 3	
	Allocated tents (x_{trs})	Utility of tents (u_{rs})	Allocated tents (x_{trs})	Utility of tents (u_{rs})
1	2	64.31	2	64.31
3	5	110.35	5	110.35
6	7	115.05	5	190.36
7	10	96.08	5	230.36
10	11	85.44	5	300.56
12	15	80.12	5	330.20

and $d = -0.0115$, u_{rs}, which is the utility of tents on different days, are computed as shown in Table 2.10. As expected, utility of tents decreases after a certain point if tents are supplied continuously at periodic intervals. Next, a scenario is considered where allocation of tents is stopped after day 3 and u_{rs} values are computed again and stored in Table 2.10. From the table, it is observed that the utility of tents gradually increases over time, leading to starvation for tents at the shelter. The utility curve for both the cases is shown in Fig. 2.9. Concave pattern of the utility curve prevails for continuous allocation.

Enumerating Utilities for Week 15

The derived utility function (5) along with the resource allocation data, provided in Table 2.2 is used to enumerate the utility for the resources Rice (R_1), Sugar (R_2) and Biscuits (R_3) at shelters of Kathmandu, Nuwakot and Sindhupalchowk for week 15. These utilities, reduced to a ten-point scale, are shown in Table 2.11. The enumerated utilities are then communicated to the control station using the opportunistic knowledge sharing scheme (discussed in the next chapter), where they are validated and used for resource planning.

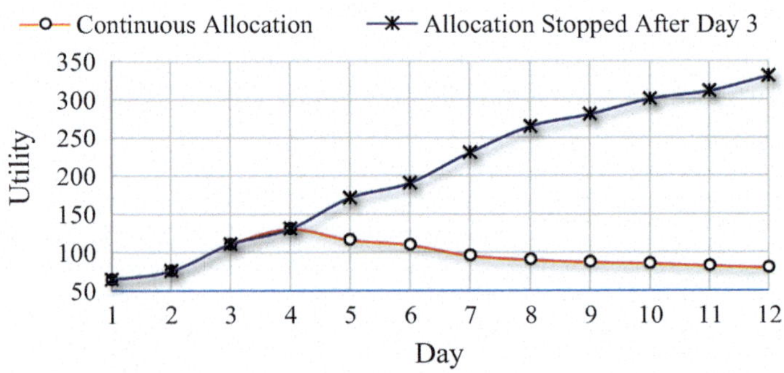

Fig. 2.9 Utility of tents against time

Table 2.11 Enumerated resource utilities for week 15

Resource	Utility in shelters		
	Kathmandu	Sindhupalchowk	Nuwakot
R_1	10	9	8
R_2	9	7	6
R_3	6	4	3

In this work, a utility function is derived and used for enumerating the utility of different emergency resources at the shelters. The function expresses utility of a resource in terms of the amount of that resource allocated at different time points. Utility of an emergency resource is its urgency at a particular time instance. It is observed that importance/urgency of a resource in a shelter decreases with time if the resource is allocated to that shelter periodically; urgency increases if the resource is not allocated after a certain time and that leads to starvation for that resource. The derived function is used to calculate the utility for emergency resources at different shelters which are then communicated to the control station where they are validated and used for resource planning.

2.5 Conclusion

In this chapter, first a PCR model is derived for predicting the exact demand for different emergency resources in a shelter, based on the constantly changing situational parameters that influence the demands. Next, a utility function is proposed that calculates the exact utility of each emergency resource in a shelter. During relief operations, the recent values of the influencing situational parameters for different emergency resources in a shelter are uploaded to the shelter-node (a laptop or a workstation in a shelter), installed therein, by the emergency supervisor present on site. The shelter-node uses the forecasting model derived in Eq. (2.1) to estimate the future demand of each of the emergency resources in that shelter. Likewise, the periodically allocated amounts of different emergency resources in a shelter are also uploaded to the shelter-node by the emergency supervisors. The shelter-node uses the utility function derived in (2.5) to estimate the current utilities of each of the emergency resources in that shelter. Thus, the demand forecast model and the utility function assist in accurate assessment for emergency resource needs at different shelters. The next obvious step would be transmitting these assessed needs to the control station for necessary resource planning.

References

1. Guo J, Zhou G (2011) Research on emergency material demand forecast method under large-scale earthquake. Value Eng 2011(22):27–29
2. Spencer S et al (2009) A multivariate time series approach to modeling and forecasting demand in the emergency department. J Biomed Inform 42(1):123–139
3. Meng C (2007) The research on emergency supplies inventory management based on fuzzy evaluation and gray neural network. Master's Thesis, Wuhan University of Technology, Hubei Sheng, China
4. Sheu J (2007) An emergency logistics distribution approach for quick response to urgent relief demand in disasters. Transp Res Part E Logistics Transp Rev 43(6):687–709
5. Sun B, Maa W, Zhao H (2013) A fuzzy rough set approach to emergency material demand prediction over two universes. Appl Math Model 37(11–12):7062–7070
6. Zhang H, Xu J (2010) Research on Emergency Material Demand Forecasting Model in Disaster Based on MLR–CBR. In: Proceedings of international conference of logistics engineering and management 2010, pp 387–404
7. Xiao W, Ya-ming Z (2011) Forecasting model of unconventional emergence incident's resource demand based on case-based reasoning. J. Xidian Univ (Social Science Edition) 2010(4)
8. Yu CH (2011) Principal component regression as a countermeasure against collinearity. In: Proceedings of Western SAS software users conference 2011, pp 1–7
9. Massy WF (1965) Principal components regression in exploratory statistical research. J Am Stat Assoc 60(309):234–256
10. Haque MM, Rahman A, Hagare D, Kibria G (2013) Principal component regression analysis in water demand forecasting. J Hydrol Environ Res 1(1):49–59
11. Saravanan S, Kannan S, Thangaraj C (2012) India's electricity demand forecast using regression analysis and artificial neural networks based on principal components. ICTACT J Soft Comput 2(4):365–370
12. Ismail NA, Abdullah SM (2016) Principal component regression with artificial neural network to improve prediction of in electricity demand. Int Arab J Inf Technol 13(1A):196–202
13. Rajab JM, MatJafri M, Lim H (2013) Combining multiple regression and principal component analysis for accurate predictions for column ozone in Peninsular Malaysia. Atmos Environ 71:36–43
14. Sousa S, Martins F, Alvim-Ferraz M, Pereira M (2007) Multiple linear regression and artificial neural networks based on principal components to predict ozone concentrations. Environ Model Softw 22(1):97–103
15. Florez JV, Lauras M, Dupont L, Charles A (2013) Towards a demand forecast methodology for recurrent disasters. WIT Trans Built Environ 133:99–110
16. Basu S, Roy S, DasBit S (2018) A post disaster demand forecasting system using principal component regression analysis and case-based reasoning over smartphone based DTN. IEEE Trans Eng Manage https://doi.org/10.1109/tem.2018.2794146
17. Guidelines for Estimating Food and Nutritional Needs in Emergencies (1997) World food programme. Available via http://www.unhcr.org/publications/operations/3b9cbef7a/wfpunhcr-guidelines-estimating-food-nutritional-needs-emergencies.html. Accessed July 2016
18. Food and Nutrition in Disasters Guidelines (2014) PAHO/WHO INCAP. Available via http://www.paho.org/disasters/index.php?option=com_content&view=article&id=553%3Anutrition-and-food-safety-in-emergency-situations-incap&Itemid=663&lang=en
19. Rapid Joint Needs Assessment Format (2014) Inter Agency Group. Available via http://iagodisha.org.in/. Accessed July 2016
20. Emergency Food Security Assessment Handbook (2005) World food programme. Available via http://documents.wfp.org/stellent/groups/public/documents/manual_guide_proced/wfp142691.pdf. Accessed July 2016
21. RAPID Needs Assessment Format (2016) Available via https://sphereindiablog.files.wordpress.com/2016/08/jrna-village-tool-revised1.pdf. Accessed April 2015

22. The Sphere Project (2015). Available via http://www.sphereproject.org/. Accessed April 2015
23. Benfield A (2015) Nepal earthquake event recap report. Available via http://thoughtleadership.aonbenfield.com/documents/201509-nepal-earthquake.pdf. Accessed July 2016
24. Google Map of Water, Food, Shelter and Medical Resources for Nepal Earthquake (2015) Available via https://www.google.com/maps/d/viewer?mid=1Iv7GILViqyJAFn5o5hi1F2Fg8mc&hl=en_US. Accessed July 2016
25. Nepal Disaster Risk Reduction Portal, Government of Nepal (2015) Available via http://drrportal.gov.np/distributed_country. Accessed July 2016
26. Nepal Earthquake Recovery Monitoring Assessment (2015) Shelter Cluster Nepal. https://www.sheltercluster.org/sites/default/files/docs/reach_npl_report_shelter_recovery_monitoring_assessment_nov2015.pdf. Accessed July 2016
27. XLSTAT (2015) Statistical software for Microsoft Excel. Available via https://www.xlstat.com/en/. Accessed July 2016
28. Basu S, Roy S, Bandyopadhyay S, DasBit S (2018) A utility driven post disaster emergency resource allocation system using DTN. IEEE Trans Syst Man Cybern Syst. https://doi.org/10.1109/tsmc.2018.2813008
29. Friedman DD (1986) The consumer: marginal value, marginal utility, and consumer surplus. In: Price theory: an intermediate text, South-Western Publishing Co (Chapter 4)
30. Krugman P, Wells R (2009) The rational consumer. In: Microeconomics. Worth Publishers, pp 269–290
31. Utility of Clams (2016) Available via http://sandovalhernandezj.people.cofc.edu/index_files/ch10.pdf. Accessed Jan 2016
32. Kuo WH, Liao W (2005) Utility-based optimal resource allocation in wireless networks. In: Proceedings of GLOBECOM 2005, pp 3408–3512
33. McConnell C, Brue S, Flynn S (2014) Law of diminishing marginal utility. In: Economics: principles, problems, & policies. McGraw-Hill Series in Economics, pp 134–153
34. Gershuny J (2009) activities, durations and the empirical estimation of utility. In: Sociology Working Papers, University of Oxford
35. Total Utility vs. Marginal Utility. Available via http://www.yourarticlelibrary.com/managerial-economics/total-utility-vs-marginal-utility-explained-with-diagram/28383/. Accessed Jan 2016
36. Rittenberg L, Tregarthen T (2016) The concept of utility. In: Principles of microeconomics, v. 1.0. Available via http://catalog.flatworldknowledge.com/bookhub/21?e=rittenberg-ch07_s02. Accessed Jan 2016
37. Total Utility (2016) AmosWEB Encyclonomic WEBopedia. Available via http://www.AmosWEB.com. Accessed Jan 2016
38. Hayes A (2016) Economics basics: utility. Investopedia Academy. Available via http://www.investopedia.com/university/economics/economics5.asp. Accessed Jan 2016

Chapter 3
Need Accumulation Over DTN

The demand forecast model and the utility function, derived in Chap. 2, are used to assess the exact need for emergency resources at different shelters. As mentioned earlier, these assessed needs are required to be transmitted and accumulated at the control station for maintaining a fully functional relief chain. In absence of stable communication infrastructure during disasters, it becomes difficult to collect the resource needs from remote and inaccessible shelters and transmit them to the control station. These result in high perception deviation at the control station about the actual needs, leading to inappropriate resource planning. This calls for designing a need accumulation scheme that would collect and transmit the shelter resource needs to the control station, even in absence of stable communication infrastructure.

In this chapter, an opportunistic knowledge sharing scheme [1] is proposed for transmitting and accumulating the shelter resource needs at the control station using a smartphone-based delay-tolerant network (DTN). This work is an extension of the "opportunistic knowledge injection" scheme [2], based on human mobility characteristics to develop a coherent global view of concentration of volunteers and relief workers in the network. As discussed in Chap. 1, the networking research community has strongly proposed the use of DTN for setting up an emergency post-disaster communication network [3–7]. The increasing availability of wireless devices, such as smartphones and their integration with technologies like Wi-Fi Direct, Bluetooth, can be harnessed to form of a DTN during or after a disaster. In this work, assuming an underlying DTN leveraged smartphone (carried by volunteers and relief workers)-based communication network, an opportunistic knowledge sharing scheme is proposed. The scheme is used for transmitting the assessed resource needs to the control station in absence of normal communication facilities. The opportunistic knowledge sharing scheme runs on the smartphones and uses dominant human mobility characteristics to collect and exchange emergency resource needs at different shelters. Thus, the scheme enables *need accumulation* which in turn reduces the gap in perception about the actual needs at different shelters.

S. Basu et al., *Reliable Post Disaster Services over Smartphone Based DTN*, Smart Innovation, Systems and Technologies 137,
https://doi.org/10.1007/978-981-13-6573-7_3

3.1 Literature Review

In this section, works on human mobility characteristics and post-disaster volunteer movements that are relevant to the proposed opportunistic knowledge sharing scheme are reviewed.

3.1.1 Human Mobility Characteristics

Researchers have proposed that the mobile devices, like smartphones, carried by volunteers may be used to set up a DTN for collection of critical information in a communication–disrupted disaster environment. However, as suggested by Wang et al. in [8], mobility of these mobile devices is mainly controlled by their carriers, the human beings, and the mobility of human beings is driven by their sociality, which is stable in the long term. Wang et al. introduce two important human mobility characteristics [9], which are obtained by analyzing the CRAWDAD trace set obtained from Dartmouth College's WLAN traces [10]. These characteristics are (i) spatial regularity and (ii) spatial locality. Spatial regularity property of human mobility indicates that people generally have a tendency of returning to a few frequently visited places, while spatial locality suggests that they usually move within a local region. Results furnished in Wang et al. in [9] and Henderson et al. in [10] suggest that the mobile nodes are found at their first two preferred locations with a probability over 70%, suggesting spatial locality. Gonzalez et al. in [11] suggest analogous human mobility characteristics that are observed from real-world data sets obtained by tracking the trajectories of 100,000 anonymous mobile phone users over a period of six months. The trace data presented by Song et al. in [12] also verify the spatial regularity and spatial locality of human mobility.

3.1.2 Post-disaster Volunteer Movement

The above study establishes spatial locality and spatial regularity as two stable human mobility characteristics. Next, some works on the mobility pattern of volunteers and relief workers in a post-disaster scenario are examined.

Yusuf Uddin et al. in [13] deal with the impact of a disaster on the movement of population and other moving agents like relief vehicles. They identify that rescue workers deployed at a neighborhood mostly work in that neighborhood and periodically return to the relief camps or some other points of interest for special instructions or in search of relief materials. Such movements of the volunteers and rescue workers exhibit the spatial locality and spatial regularity characteristics of human mobility, explained in [9, 10].

Nelson et al. in [14] propose a disaster mobility model in which an object's movement pattern is caused by environmental events. Among other objects, authors define a special object, the rescue workers, that first approach an event and then, upon reaching the event, travel to a predefined location. This movement pattern is then repeated continuously and hence shows evidence of spatial regularity.

Aschenbruck et al. in [15] and [16] provide a model that realistically represents the movements in a disaster area scenario. The model is based on an analysis of tactical issues of civil protection. In this work, authors propose that rescue teams and fire brigades do not walk around randomly; rather, the movements are driven by tactical reasons. Such movements are periodic in nature that repeats itself regularly.

Bai et al. in [17] discuss a reference point group mobility (RPGM) model that can be used during disaster relief where various rescue crews (e.g., firemen, policemen and medical assistants) form different groups and work cooperatively. In the RPGM model, each group has a center, which is either a logical center or a physical centre. Group members, randomly placed in the neighborhood, work there and return to the centre at periodic intervals.

The above study confirms that the mobility pattern of volunteers and relief workers during post-disaster relief operations is governed by spatial locality and spatial regularity. These stable human mobility characteristics are exploited for devising the opportunistic knowledge sharing scheme for transmitting emergency resource needs to the control station over a DTN leveraged smartphone (carried by volunteers and relief workers)-based post-disaster communication network.

3.2 System Model

In this section, the network architecture, distribution of volunteers to different regions and the interaction pattern of the volunteers are described.

3.2.1 Network Architecture

In the proposed work, it is assumed that the smartphones carried by the volunteers communicate over a DTN and create a post-disaster communication network. Such a DTN-based post-disaster communication network comprises of three types of nodes—shelter-node, control-node and forwarder-node [2]. The roles of these nodes are summarized below. A typical post-disaster communication network is illustrated in Fig. 3.1.

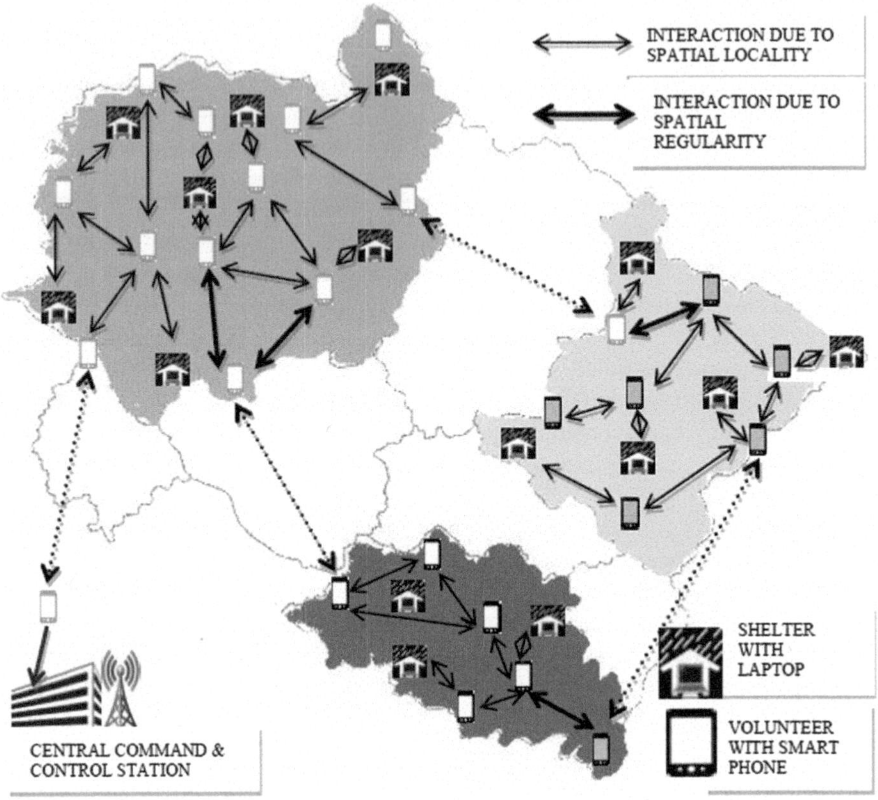

Fig. 3.1 A post-disaster communication network showing interaction pattern among volunteers

3.2.1.1 Shelter-node

A *shelter-node* is a laptop or workstation in the shelter. Situational parameters of a shelter like number of victims, the type and degree of emergency they have, their consumption pattern, rate of resource replenishment, rate of migration are uploaded to the shelter-node, installed therein, by the emergency managers present on site. The shelter-node dynamically forecasts the demand for different emergency resources using the proposed PCR model and enumerates the utility of such resources using the derived utility function. The shelter-node in a shelter periodically broadcasts these demands and utilities towards the forwarder-nodes that are working in and around that shelter.

3.2.1.2 Forwarder-node

Forwarder-nodes are the volunteers carrying DTN-enabled smartphones. These nodes work around a shelter and receive the resource needs of that shelter broadcast by the shelter-node therein. These nodes opportunistically interact with each other due to their mobility; automatically exchange resource needs of the shelters; build a global view of need information; and eventually transmit them to the control station.

Whenever these nodes move into the radio range of each other, communication links are established. Information exchange between the DTN-enabled smartphones carried by the volunteers is accomplished by using their Bluetooth or Wi-Fi-Direct interfaces, whenever two such devices come within the radio range of each other. Information is forwarded in multiple hops on a peer-to-peer fashion using DTN routing protocols. Such Bluetooth or Wi-Fi-Direct-based transmission eliminates the necessity of traditional communication requirements like GSM, GPRS or Wi-Fi for collecting and exchanging need information across the network. The forwarder-nodes are an essential part of any DTN as routing can be implemented only with help of these nodes. All such nodes are assumed to be resource constrained in terms of processor and battery, and every such node forwards its own messages along with messages of other nodes.

3.2.1.3 Control-Node

The *control-node* (a workstation or a server at the control station) receives updated resource needs from the forwarder-nodes; carries out a case-based reasoning driven technique to validate/estimate the needs; and executes a utility driven optimal resource allocation strategy to allocate resources to shelters targeting minimization of resource deficit and deployment time.

3.2.2 Distribution of Volunteers

During and after disaster, victims normally take shelter in nearby safe areas like school buildings, temporary tents in some highland areas and other risk-free zones. Thus, shelters get localized in specific parts of a disaster affected area. To represent such localities, the entire affected area is virtually divided into non-intersecting neighborhoods, called regions Z. These regions may be of any shape and size and may be adjacent or geographically separated. As discussed in Sect. 3.1.1, spatial locality property of human mobility indicates that people usually move within a local region, while spatial regularity suggests that they have a tendency of returning to a few frequently visited places. Due to the spatial locality, it is normal for a volunteer to be present in a particular region (designated as its home region) more often than in other regions. Therefore, each forwarder-node is allocated to serve a particular shelter in its home region. It is assumed that each forwarder-node, FN is aware of its own physical location at any point of time and a mapping function can be used to determine the region in which it is present at that point of time. Furthermore, since post-disaster relief operations are event-driven and there might be a need for additional volunteers in a shelter, a volunteer may have to visit and serve shelters in other regions whenever an emergency arises therein. Thus, a forwarder-node, owing to spatial regularity, periodically visits other regions.

3.2.3 *Interaction Pattern of Volunteers*

Due to spatial locality of volunteers, they interact with other volunteers working in their locality more frequently. If any volunteer is required to visit a remote location, she is likely to return to her originally assigned location eventually due to spatial regularity. Therefore, in the present context, the interaction pattern of volunteers exhibits spatial locality as well as spatial regularity. Such interactions are shown in Fig. 3.1. Interactions among volunteers (or forwarder-nodes) carrying smartphones, of the same region, due to spatial locality, are denoted by thin arrows. Interactions among volunteers of remote regions, due to spatial regularity, are denoted by bold arrows. Dotted arrows signify volunteer movements. Such movements and interaction patterns are used in our opportunistic knowledge sharing scheme discussed below.

3.3 Opportunistic Knowledge Sharing Scheme

In this section, an opportunistic knowledge sharing scheme leveraged by human mobility characteristics is proposed that transmits and accumulates emergency resource needs of far-flung shelters to the control station over a DTN-based communication architecture. In this scheme, a volunteer periodically captures knowledge about resource needs in different shelters and opportunistically exchanges this knowledge with other volunteers whom she meets due to mobility. Such exchanges are implemented with the help of Bluetooth or Wi-Fi Direct interface of the forwarder-nodes (DTN-enabled smartphones, carried by the volunteers) following DTN routing protocols [18–26].

Knowledge, in the proposed scheme, refers to the emergency resource needs of a shelter. Need, in turn, refers to utility as well as demand for resources, which evolve dynamically depending on the parameters like number of victims at a particular time, the type and degree of emergency they have, their consumption pattern, rate of resource replenishment and rate of migration. These parameters in a shelter are uploaded to the shelter-node, installed therein, by the emergency managers present on site. The shelter-node dynamically forecasts the demand for different emergency resources using the proposed PCR model and enumerates the utility of such resources using the derived utility function. The shelter-node in a shelter periodically broad-

Table 3.1 (a) Demand vector for shelter S_i (b) utility vector for shelter S_i

Shelter	Recency	Demand for resource type			
		1	2	...	R
(a)					
S_i	$t_{(k,i)}$	d_{1i}	d_{2i}	...	d_{Ri}

Shelter	Recency	Utility for resource type			
		1	2	...	R
(b)					
S_i	$t_{(k,i)}$	u_{1i}	u_{2i}	...	u_{Ri}

casts this knowledge towards the forwarder-nodes that are working in and around that shelter. Different phases of the opportunistic knowledge sharing scheme are described below.

3.3.1 Local Knowledge Acquisition

In this phase, a forwarder-node *FN* serving a shelter S_i in region Z_k periodically captures the demands of different resources in that shelter and stores them in a list called demand vector, as shown in Table 3.1a, where $t_{(k,i)}$ denotes recency of the vector. It must be noted that $t_{(k,i)}$ is a local timestamp generated by S_i and not a global timestamp and hence does not assume synchronized clocks across the affected area. d_{rs} denotes demand for the rth resource at sth shelter. Since the shelter-node periodically broadcasts updated utilities, *FN*, after every predetermined time interval δ_t, updates its demand vector with most recent utilities captured from the shelter-node. Similarly, *FN* periodically records the resource utilities in S_i in a time stamped list called utility vector, shown in Table 3.1b. Thus, the demand and utility vectors for S_i contain the most updated demands and utilities of resources and reflect the dynamically changing resource needs therein.

3.3.2 Local Knowledge Sharing and Regional Knowledge Acquisition

In addition to demand vector, *FN* maintains a demand matrix, shown in Table 3.2a, to record demands for emergency resources in each shelter in its home region Z_k. Initially, this matrix has only one row that consists of resource demands of its own shelter S_i. When *FN* meets another node FN', working in shelter S_j, belonging to the same region Z_k, due to spatial locality, they exchange demand vectors of their respective shelters and also exchange demand vectors of shelters other than S_i and S_j, available with them. Forwarder-nodes use these collected demand vectors for

Table 3.2 (a) Demand matrix for shelter S_i, (b) utility matrix for shelter Si

Shelter	Recency	Demand for resource type			
		1	2	…	R
(a)					
S_1	$t_{(k,1)}$	d_{11}	d_{21}	…	d_{R1}
…	…	…	…	…	…
S_i	$t_{(k,i)}$	d_{1i}	d_{2i}	…	d_{Ri}
…	…	…	…	…	…
S_P	$t_{(k,P)}$	d_{1P}	d_{2P}	…	d_{RP}
Shelter	Recency	Utility for resource type			
		1	2	…	R
(b)					
S_1	$t_{(k,1)}$	u_{11}	u_{21}	…	u_{R1}
…	…	…	…	…	…
S_i	$t_{(k,i)}$	u_{1i}	u_{2i}	…	u_{Ri}
…	…	…	…	…	…
S_P	$t_{(k,P)}$	u_{1P}	u_{2P}	…	u_{RP}

adding new rows to their respective demand matrices to populate them with resource demands of other shelters in that region. This way, all demand matrices record almost similar and recent resource demands in the shelters of a region. However, *FN* while replacing an existing row, corresponding to a shelter S_l, with a demand vector received from FN', checks the recency, $t_{(k,l)}^{received}$, of the received vector. It uses the received demand vector to replace an existing row in its demand matrix only if

$$t_{(k,l)}^{received} - t_{(k,l)}^{exixting} > 0$$

If *FN* already has a row in its demand matrix corresponding to shelter S_j (provided by some other forwarder-node), it unconditionally replaces that row with the demand vector of S_j received from FN', because FN' being a volunteer of S_j, provides the most authentic demand vector of S_j. Likewise, *FN* builds a utility matrix, shown in Table 3.2b, revealing the dynamically changing resource utilities in shelters of its home region.

3.3.3 Regional Knowledge Sharing and Global Knowledge Acquisition

Each forwarder-node in Z_k would have analogous versions of the demand matrix for Z_k. Now, when *FN* from region Z_k gets the relatively rare opportunity of meeting FN'' working in Z_m, due to spatial regularity, they exchange their regional demand

Table 3.3 Comprehensive overhead analysis of opportunistic knowledge sharing technique

Phases	Storage overhead (bytes)	Computation overhead (instructions)	Communication overhead (bytes)	
			Transmission	Reception
Local knowledge acquisition	$2kR$	$2R$	0	$2kR$
Local knowledge sharing and regional knowledge acquisition	$2kPR$	$2(N-1)(P-2)$	$2k(N-1)R(P-2)$	$2k(N-1)(P-2)R$
Regional knowledge sharing and global knowledge acquisition	$2kPQR$	$2NP(Q-2)(Q-1)$	$2kNP(Q-2)(Q-1)R$	$2kNP(Q-2)(Q-1)R$
Global knowledge sharing	0	$2(N-1)P(Q-1)$	$2k(N-1)P(Q-1)R$	0
Total	$2kR(1+P+PQ)$	$2(NPQ^2-2NPQ+2NP-2N-PQ+R+2)$	$2kR(NPQ^2-2NPQ+2NP-P-2N+2)$	$2kR(NPQ^2-3NPQ+3NP-2N-P+3)$

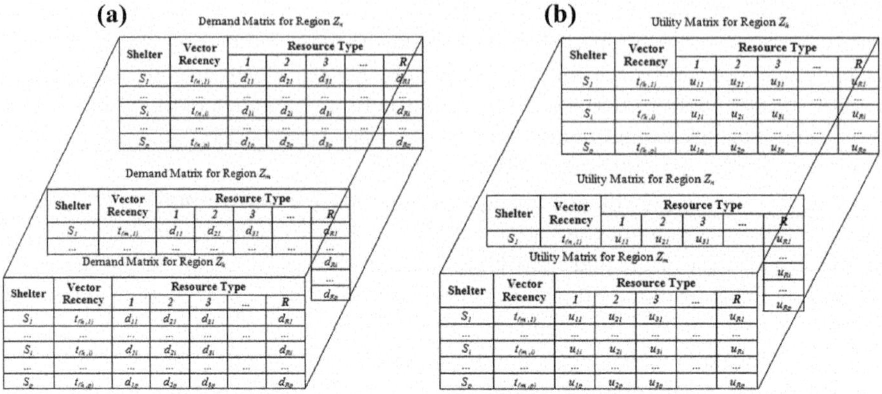

Fig. 3.2 **a** Demand set, **b** utility set

matrices and also gather demand matrices of regions other than Z_k and Z_m. These newly collected demand matrices are then collated to form a demand set, as shown in Fig. 3.2a, to record the demands for emergency resources in shelters belonging to different regions. The demand set of *FN* initially contains only one demand matrix pertaining to its home region Z_k and incrementally gets populated with demand matrices of other regions like Z_m. This way, all demand sets record almost similar resource demands in shelters of different regions, and thus, a demand set contains a global view of resource needs in the entire disaster affected area. However, *FN* while updating an existing demand matrix, corresponding to a region Z_n, with a demand matrix received from FN'', checks the recency, $t^{received}_{(n,\bullet)}$, of each row of the received matrix. *FN* uses a row in the received demand matrix to replace the corresponding row in the existing matrix in its demand set only if

$$t^{received}_{(n,\bullet)} - t^{existing}_{(n,\bullet)} > 0$$

This way, a demand set always holds the most updated resource demands of shelters in different regions. *FN* copies new rows that were not there in its existing demand matrix, from the received demand matrix. If *FN* already has a demand matrix corresponding to Z_m in its demand set (provided by some other forwarder-node), it unconditionally replaces that matrix with the one received from FN'', because FN'' being a volunteer of Z_m provides the most authentic demands of Z_m. The demand set of *FN* initially contains only one demand matrix pertaining to its home region Z_k and incrementally gets populated with demand matrices of other regions like Z_m. Thus, demand sets of *FN* and FN'' record almost similar resource demands in shelters of different regions, and thus, these sets contain resource needs of the entire disaster affected area. *FN* follows similar procedures to build the utility set as shown in Fig. 3.2b.

3.3.4 Global Knowledge Sharing

On returning to its home region Z_k due to spatial regularity, *FN* shares its newly acquired demand and utility matrices, pertaining to other regions, with other forwarder-nodes in its home region, thus disseminating the knowledge further. All forwarder-nodes in this region update their demand and utility sets with these newly acquired matrices so as to keep the current resource needs of the shelters of other regions. Thus, knowledge sharing enables all forwarder-nodes in the network to stay abreast of the current resource needs in all shelters without the having to meet all other nodes in the network. In this process, a single interaction between one volunteer from each region can provide a reasonably accurate view resource needs of the entire affected area. When any one of the forwarder-nodes manages to visit the control station (located outside the affected area), it delivers its own copy of demand and utility set therein. These sets have the resource needs not only of her own shelter but the needs of the entire disaster affected area.

3.3.5 Algorithm

This section presents an algorithm for the opportunistic knowledge sharing scheme discussed above. The algorithm presented below runs on the DTN-enabled smartphones carried by the volunteers (forwarder-nodes), where data exchange (knowledge sharing) occurs through the Bluetooth or Wi-Fi direct interface of the smartphones, following DTN routing protocols. The algorithm carries out an in-network processing of collected information and dynamically builds a global picture of resource needs in the entire affected area, which is consistent for all forwarder-nodes in the network. The algorithm is presented for collecting and exchanging resource demands. The same algorithm is used for collecting and exchanging resource utilities as well.

Algorithm : Opportunistic Knowledge Sharing

Input : Demand Vector containing old utilities of *FN*'s shelter, Demand Matrix containing Demand Vector of *FN*'s shelter and Demand Set containing Demand Matrix of *FN*'s own region

Output : Demand Vector containing recent utilities of *FN*'s shelter, Demand Matrix containing Demand Vectors of different shelters and Demand Set containing Demand Matrices of remote regions

Runs At : Forwarder-node (*FN*, DTN enabled smartphone, carried by volunteers)

```
do
  for each FN ∈ Z_k do                                   // FN works in shelter S_i in region Z_k
    if t^{current} − t^{FN}_{(k,i)} ≥ δ_t then           // t^{current}: current time, t^{FN}_{(k,i)}: recency of FNs Demand Vector
      t^{FN}_{(k,i)} := t^{current}                      // set t^{FN}_{(k,i)} to current time
      for each r in 1, 2, ... R do                       // considering R resource types
        DV^{t^{FN}_{(k,i)}}_{(k,i)}[r] := current demand // DV^{t^{FN}_{(k,i)}}_{(k,i)}[r]: demand of r^{th} resource in Demand Vector of S_i
      end for
    end if
    for each FN' ∈ Z_m do                                // FN' works in region Z_m
      if Z_k ≠ Z_m then                                  // FN and FN' have different home regions
        DM^{FN}_m := DM^{FN'}_m                          // DM^{FN}_m : Demand Matrix of Z_m in Demand Set of FN
        for each DM^{FN'}_n in DS_{FN'} do               // DS_{FN'} : Demand Set of FN'
          for each i in 1, 2, ... P do                   // Assuming P shelters in Z_n
            if t^{FN'}_{(n,i)} > t^{FN}_{(n,i)} then     // FN' holds a more recent Demand Vector in DM^{FN'}_n
              DV^{t^{FN}_{(n,i)}}_{(n,i)} := DV^{t^{FN'}_{(n,i)}}_{(n,i)}   // replace FNs Demand Matrix of Z_n with FN's version
            end if
          end for
        end for
      else                                               // FN' works in S_j in FNs home region Z_k
        DV^{t^{FN}_{(n,j)}}_{(k,j)} := DV^{t^{FN'}_{(n,j)}}_{(k,j)}         // DV^{t^{FN}_{(n,j)}}_{(k,j)} : Demand Vector of S_j, in DM^{FN}_k
        for each l in 1, 2, ... P do                     // l ≠ i and l ≠ j
          if t^{FN'}_{(k,l)} > t^{FN}_{(k,l)} then       // FN' holds a more recenet Demand Vector of S_l
            DV^{t^{FN}_{(k,l)}}_{(k,l)} := DV^{t^{FN'}_{(k,l)}}_{(k,l)}     // replace FNs Demand Vector of S_l with FN's version
          end if
        end for
        for each DM^{FN}_n in DS_{FN}                    // DS_{FN} : Demand Set of FN
          for each i in 1, 2, ... P do                   // DM^{FN}_n: Demand Matrix of Z_n in US_{FN}
            if t^{FN}_{(n,i)} > t^{FN'}_{(n,i)} then     // knowledge sharing, where newly acquired demand
              DV^{t^{FN'}_{(n,i)}}_{(n,i)} := DV^{t^{FN}_{(n,i)}}_{(n,i)}   // matrices of FN are shared with forwarder − nodes
            end if                                       // FN' having same home region as FN
          end for
        end for
      end if
    end for
  end for
while W ≠ ∅                                              // W : set of all forwarder − nodes
```

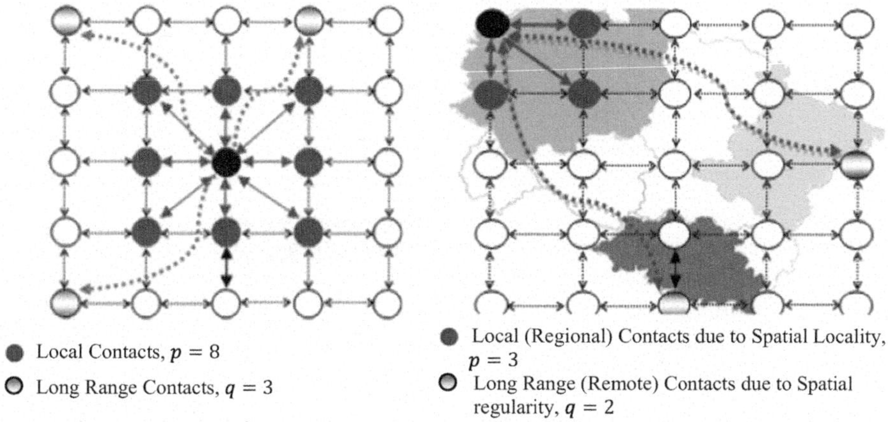

(a) Grid-based representation

(b) Application to post disaster communication network

Fig. 3.3 Small world network

3.3.6 Expected Generation Time for Global View of Resource Needs

In this section, the expected time for generating the global view of resource needs in the entire affected area is figured out. Recent studies indicate that the small world phenomenon [27], initially identified for social networks, is also useful to model human mobility-based DTNs and successful routing protocols in such DTNs can be designed by exploiting the properties of a small world network [28]. This acts as a motivation to utilize expressions for expected message delivery time in a small world network to find the expected generation time of the global view of resource needs in our human mobility-based DTN. Apart from other important properties, a small world network has two fundamental features—"short average path length" and "high cluster coefficient" [27]. "Short average path length" means two people are connected by a short chain of acquaintances, and "high cluster coefficient" denotes people are likely to make friends with their friends' friends.

Kleinberg [29] represents a small world network as a two-dimensional grid with directed edges having several local connections and a few long-range connections. Such a representation with n^2 nodes arranged in a $n \times n$ grid is shown in Fig. 3.3a. Here, p is the range of local contacts and q is the number of long-range contacts. Our DTN-based post-disaster communication network characterized by spatial locality and regularity can also be represented as a grid-based small world network with local and long-range connections, as shown in Fig. 3.3b.

Kleinberg uses the fundamental properties of a small world network to derive its expected message delivery time $E(T_M)$ *as:*

$$E(T_M) = \begin{cases} \Omega\big(n^{(2-r)/3}\big) & \text{if } 0 \le r < 2 \\ O(\log n)^2 & \text{if } r = 2 \\ \Omega\big(n^{(r-2)/(r-1)}\big) & \text{if } r > 2 \end{cases}$$

where r is average clustering coefficient of the network. The average clustering coefficient of a DTN, with different mobility models, is shown to lie between 0 and 1 [30], i.e., $E(T_M) = \Omega\big(n^{(2-r)/3}\big)$ for a DTN.

In the regional knowledge acquisition phase of the proposed opportunistic knowledge sharing scheme, a forwarder-node populates its demand and utility matrices with demand and utility vectors of all the shelters in its own region. In the best case, a forwarder-node can collect demand and utility vectors of all shelters in its region from a single node working in that region. On the other hand, in the worst case, the forwarder-node has to collect demand and utility vectors of each shelter from a separate node. Assuming P shelters in a region, there are $2 \times P$ messages (demand and utility vectors) transfers, and hence, in the worst case, the expected generation time of a demand matrix is $2 \times P \times \Omega\big(n^{(2-r)/3}\big)$. In the global knowledge acquisition phase, a forwarder-node populates its demand and utility sets with demand and utility matrices of all the regions. Similar to the regional knowledge acquisition phase, in the worst case, the forwarder-node has to collect demand and utility matrices of each region from a separate node. Assuming Q regions, there are Q messages (demand and utility matrices) transfers and the expected generation time of a demand set is $2 \times Q \times \Omega\big(n^{(2-r)/3}\big)$. Therefore, in the worst case, the expected generation time of a global view of resource demands in the entire affected area sums up to:

$$E(T_G) = 2 \times (P + Q) \times \Omega\left(n^{(2-r)/3}\right)$$

$E(T_G)$ lies between $2 \times (P + Q) \times \Omega\big(n^{1/3}\big)$ and $2 \times (P + Q) \times \Omega\big(n^{2/3}\big)$ for r $\in [0, 1]$. Thus, the proposed opportunistic knowledge sharing scheme builds a global view of resource needs in at most linear time to reduce the gap in perception about actual needs.

3.4 Performance Analysis

The effectiveness of the proposed opportunistic knowledge sharing scheme reported in the earlier section is evaluated through both qualitative and quantitative analyses.

3.4.1 *Qualitative Analysis*

This section presents the performance of the scheme in terms of storage, computation and communication overheads. As per the system model, specified in Sect. 3.2, shelter-nodes and the control-node are high-end computing devices like laptop, workstation and server. On the other hand, the forwarder-nodes that run the opportunistic knowledge sharing algorithm are DTN-enabled smartphones, which are constrained in terms of processor, battery and storage. Therefore, the storage, computation and communication requirements of this algorithm are enumerated to evaluate its suitability for resource-constrained smartphones. The above overheads are computed for an operational period, which is typically 12 h [31] for our application. In particular, the overheads are computed for the worst case scenario that occurs if a particular forwarder-node executes all four phases of the opportunistic knowledge sharing algorithm and meets all other nodes in the network.

We assume Q regions, N forwarder-nodes per region, P shelters and R resource types. In the local knowledge sharing phase, a forwarder-node interacts with a maximum of $(N-1)$ other nodes working in its own region. In the regional knowledge sharing phase, the forwarder-node interacts with a maximum of $N(Q-1)$ other nodes working in other regions. The Samsung Galaxy Grand 2 smartphone is considered for this analysis that has a Qualcomm Snapdragon processor with ARM core. Battery capacity is 2600 mAh with a voltage of 3.8 V and initial energy of 35,568 J [32].

3.4.1.1 Storage Overhead

Storage overhead is measured in terms of the memory requirement for storing the demand and utility vectors, matrices and sets in the smartphones. Demand and utility vectors, populated at the local knowledge acquisition phase, consist of R entries each corresponding to R resources, i.e. $2R$ entries in total. Demand and utility matrices, populated at the local knowledge sharing and regional knowledge acquisition phase, have P rows for P shelters. Thus, each matrix contains PR entries, i.e. $2PR$ entries in total. Demand and utility sets, populated at the regional knowledge sharing and global knowledge acquisition phase, are composed of Qdemand matrices for Q regions. So, each of the sets has PQR entries, i.e. $2PQR$ entries in total. Assuming k bytes for each entry, storage overhead of our algorithm for a forwarder-node, worked out in Table 3.3, comes out to be:

$$2kR(1 + P + PQ) \text{ bytes}$$

3.4.1.2 Computation Overhead

Computation overhead is measured as the number of instructions required for executing different phases of the opportunistic knowledge sharing algorithm. Considering that the algorithm is run for utilities as well as for demands, the computation overhead for a forwarder-node, figured out in Table 3.3, sums up to:

Table 3.4 Required energy consumption for relevant operations (for the processor under consideration)

Operation	No. of operations	Energy consumption (J)
LOAD	$4NPQ^2 - 8NPQ + 8NP + 4NQ - 4PQ - 8N + 2R + 8$	0.225
MOV	$NPQ^2 - 2NPQ + 2NP + NQ - PQ - N + R + 3$	0.1947
STORE	$NPQ^2 - 2NPQ + 2NP + NQ - PQ - N - Q + R + 2$	0.125
CMP	$NPQ^2 - 2NPQ + 2NP + NQ - PQ - 2N + 2$	0.751
SUB	1	0.2052

$$2\left(NPQ^2 - 2NPQ + 2NP - 2N - PQ + R + 2\right) \text{ instructions}$$

Assuming all possible contacts, the computation time of the opportunistic knowledge sharing algorithm sums up to

$$\boldsymbol{T} = \max\left[R, (N-1) \times (Q-1) \times P\right], \quad \text{assuming } Q \geq 2$$

If N is assumed to be significantly large in comparison to R, which is quite pragmatic, the worst case computation time is

$$O(N \times P \times Q)$$

For practical values of N, P and Q the computation time will be considerably moderate.

The computation overhead of a forwarder-node for executing these phases is the total number of low-end operations it performs. Table 3.4 shows the required number of such low-end operations and corresponding energy requirements [33]. Using these, energy requirements and the number of each type of instruction, the computation energy overhead sums up to:

$$1.9715\left(NPQ^2 + NQ - PQ\right) + 3.943(NP - NPQ) - 3.6 \\ + O.7701R - 0.3197Q + 4.3429 \text{ J}$$

3.4.1.3 Communication Overhead

Communication overhead is computed in terms of the total number of bytes exchanged (transmitted and received) by a forwarder-node at different phases of the opportunistic knowledge sharing scheme. The communication overhead, as shown in Table 3.3, sums up to:

Transmission overhead	$2kR\left(NPQ^2 - 2NPQ + 2NP - PQ - 2N + 2\right)$ bytes
Reception overhead	$2kR\left(NPQ^2 - 3NPQ + 3NP - 2N - P + 3\right)$ bytes

The Bluetooth communication energy consumption for the smartphone under consideration is 0.384 J/s for transmitting 1 Mb data and 0.329 J/s for receiving 1 Mb data [32]. Also, Bluetooth 1.2 has an average data transfer rate of 1 Mb/s. Using the communication requirements specified in Table 3.3, the communication energy overhead sums up to:

$(1024)^{-2} \times 0.384 \times R \times k\left(NPQ^2 - 2NPQ + 2NP - PQ - 2N + R\right)$	Joules for transmission
$(1024)^{-2} \times 0.329 \times k(NPQ^2 - 3NPQ + 3NP - 2N - P + 3$	Joules for reception

3.4.1.4 Illustrative Example

Assuming 3 regions, 10 resource types, 4 shelters, 40 volunteers per region and 4 bytes for a single demand or utility value, i.e., $N = 40$, $P = 4$, $Q = 3$, $R = 10$ and $k = 4$, the worst case storage, computation, communication and energy overheads are computed as:

Storage overhead	1360 bytes $\approx$ 1.32 Kb
Computation overhead	1440 instructions
Communication overhead	
Transmission overhead	56,800 bytes $\approx$ 55.46 Kb
Reception overhead	31,620 bytes $\approx$ 31.17 Kb
Energy overhead	
Communication energy	0.0002 J
Computation energy	1797.6 J

The energy consumption for running our opportunistic knowledge scheme for the entire operational period sums up to 1797.6002 J. The initial energy of the smartphone is 35,568 J, and energy consumed for running background processes is 0.562 J/s. Therefore, our opportunistic knowledge sharing scheme can be run for at most 14 h on a smartphone dedicated to this purpose. Since volunteers in a post-disaster scenario work only during daytime, typically for 12 h a day, our scheme can be run for the full operational period without further charging.

3.4.2 *Quantitative Analysis*

This subsection provides the quantitative analysis of the scheme.

3.4.2.1 Simulation Environment and Data

The simulation environment and data used for performance analysis of the PCR-based demand forecasting model, described in Chap. 2 (Sects. 2.2.4.1 and 2.2.4.2), are used for conducting quantitative performance analysis of the opportunistic knowledge sharing scheme as well.

3.4.2.2 Simulation Set-up

The ONE simulator [18] is used to implement the opportunistic knowledge sharing scheme in a DTN environment. The map in [34] is overlaid on ONE to plot nine shelters (S_1 to S_9) in three regions—S_1 to S_3 in Kathmandu, S_4 to S_6 in Sindhupalchowk and S_7 to S_9 in Nuwakot. Ten forwarder-nodes, i.e., volunteers with smartphones, are assumed to be working in a shelter. The forwarder-nodes with node IDs 1–30 serve the Kathmandu region, 31–60 serve the Sindhupalchowk region and 61–90 serve the Nuwakot region. The forwarder-nodes with Node IDs 1–10 serving shelter S_1 interact more often with nodes with Node IDs 11–30 working in the same region. Due to sparse nature of the network, the probability of interaction of Node IDs 1–10 with distant forwarder-nodes with Node IDs 31–90 is very less. The control station is set up at a distant location on the map. The shelter-nodes and the control-node are stationary. However, the forwarder-nodes are mobile nodes and follow the post-disaster mobility (PDM) model proposed by Uddin et al. in [13]. Due to spatial locality, nodes spend 90–95% of their time in their home region and 5–10% time in other regions. Since post-disaster relief operations were usually carried out during the daytime, typically from 6 a.m. to 6 p.m. [31], the simulation time is kept as 12 h. A shelter-node updates demands and utilities, once in a day and broadcasts them towards the forwarder-nodes, who in turn update their demand and utility vectors accordingly.

The PCR forecast demands (Sect. 2.2.3 of Chap. 2) and the enumerated utilities (Sect. 2.3.1 of Chap. 2) of resources rice (R_1), sugar (R_2), and biscuits (R_3), in Chap. 2, derived by the shelter-nodes at shelters of Kathmandu, Sindhupalchowk and Nuwakot for week 15, are shown in Table 3.5a, b, respectively. It is assumed that these demands and utilities at a representative shelter in a region (Kathmandu, Sind-

Table 3.5 (a) PCR forecast resource demands for week 15, (b) enumerated resource utilities for week 15

Resource	Demand in shelters		
	Kathmandu	Sindhupalchowk	Nuwakot
(a)			
R_1	52	72	66
R_2	1	4	3
R_3	30	51	49
Resource	Utility in shelters		
	Kathmandu	Sindhupalchowk	Nuwakot
(b)			
R_1	10	9	8
R_2	9	7	6
R_3	6	4	3

Table 3.6 Parameters for simulation in one

Parameter	Value
Number of shelters	9
Number of forwarder-nodes	90
Simulation period	12 h
Node speed	10–50 km/h
Transmission range	10 m
Message size	0.5 MB
Buffer size	5 MB
Message TTL	720 min
Movement model	PDM
Routing protocol	First contact, spray and wait

hupalchowk or Nuwakot) are same for other shelters in that region. These demand and utility values are used as input to the ONE simulator for running the opportunistic knowledge sharing algorithm. The simulation is run—firstly, for forwarding single copy of demand and utility vectors across the network, using First Contact Routing [18], and secondly, for forwarding multiple copies of demand and utility vectors using spray and wait routing [21]. In the multicopy scheme, four copies of each vector are transmitted. Table 3.6 lists the parameter values used in simulation.

3.4.2.3 Simulation Metrics

- A metric called perception deviation is used for evaluating the performance of the opportunistic knowledge sharing scheme. Two definitions of perception deviation are considered—one for demands and the other for utilities, as given below:
- ***Perception Deviation of Demands*** (***PD***$_d$) $= \frac{1}{R}\sum_{r=1}^{R}\frac{\left|\widehat{D_{rs}}-D'_{rs}\right|}{D_{rs}} \times 100$
- ***Perception Deviation of Utilities*** (***PD***$_u$) $= \frac{1}{R}\sum_{r=1}^{R}\frac{\left|\widehat{U_{rs}}-U'_{rs}\right|}{U_{rs}} \times 100$

where R is the total number of resources, $\widehat{D}_{rs}$ is the PCR model forecast demand, $\widehat{U}_{rs}$ is the utility function enumerated utility, and D'_{rs} and U'_{rs} are the demand and utility delivered by a volunteer, using the proposed scheme. An ideal value of PD_d and PD_u is 0.

3.4.2.4 Simulation Results

After 12 h of simulation, a global demand matrix and a global utility matrix, shown in Table 3.7a, b, respectively, get generated by collating all demand and utility sets accumulated at the control-node during this period. Recency $t_{(k,i)}$ of each demand and utility value in the received sets is checked so that only recent values are stored. The matrices have some of the demand and utility values missing (lost due to network related issue). These matrices are used to evaluate the performance of the proposed opportunistic knowledge sharing algorithm. The simulation is performed for demand as well as utility values, and results in both cases are analogous. However, simulation results for only the demand values are presented. The simulation results presented here are the average of 20 independent runs. Three sets of experiments are performed to evaluate the efficiency of the proposed scheme. Although the simulation is run to measure perception deviation for all 12 shelters, results for only shelter S_1 provided for the first two experiments.

Opportunistic Knowledge Sharing Against Adhoc Forwarding: The opportunistic knowledge sharing scheme is compared against an ad hoc forwarding scheme in which forwarder-nodes arbitrarily exchange only local knowledge (i.e., demand and utility vectors) with all volunteers. They do not share any regional or global knowledge. The comparison is done in terms of perception deviation.
Figure 3.4 shows PD_d about demands at shelter S_1 for each forwarder-node (node IDs 1–90) considering the single-copy scheme. In case of ad hoc forwarding, as shown in Fig. 3.4a, local nodes 1–30 have a PD_d of 15–35% about the demands in S_1, and distant nodes 31–90 have around 89–97%. In case of opportunistic knowledge sharing, as shown in Fig. 3.4b, local volunteers have a PD of 14–28% and distant nodes have around 27–37%. It is evident that in case of opportunistic knowledge sharing, distant volunteers have significantly less PD_d compared to ad hoc forwarding.

Table 3.7 (a) Global demand matrix, (b) global utility matrix

Resource	Demand in shelter								
	Kathmandu			Sindhupalchowk			Nuwakot		
	S_1	S_2	S_3	S_4	S_5	S_6	S_7	S_8	S_9
(a)									
R_1	52	105	52	72	72	72	135	66	66
R_2	1	1	9	4	–	4	3	3	3
R_3	56	30	30	99	51	51	49	123	–
Resource	Utility in shelter								
	Kathmandu			Sindhupalchowk			Nuwakot		
	S_1	S_2	S_3	S_4	S_5	S_6	S_7	S_8	S_9
(b)									
R_1	10	10	–	2	9	9	8	8	8
R_2	18	9	9	7	7	–	6	–	6
R_3	6	6	6	4	14	4	3	3	3

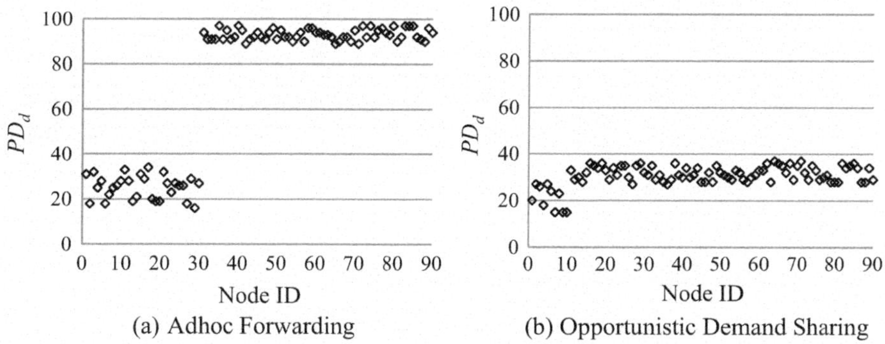

Fig. 3.4 Perception deviation about resource demands at shelter S_1 with single-copy scheme

Figure 3.5 shows PD_d about demands at shelter S_1 for each forwarder-node (node IDs 1–90) considering multicopy scheme. In case of ad hoc forwarding, as shown in Fig. 3.5a, *PD* of distant nodes using local demand sharing scheme is 86–95%. In case of opportunistic knowledge sharing, as shown in Fig. 3.5b, PD_d of all nodes is 12–21%. It is observed from the plots that in both the cases local volunteers have less PD_d about demands in the shelter S_1 compared to the distant volunteers. However, in case of opportunistic knowledge sharing, distant volunteers have significantly less PD_d compared to ad hoc forwarding. The improvement is due to the increased number of transmitted demand vectors.

Average Perception Deviation over Time: Next, the average perception deviation of local and distant nodes at intervals of 2 h, using the opportunistic knowledge sharing scheme, is calculated for shelter S_1. Figure 3.6 shows the average PD_d about

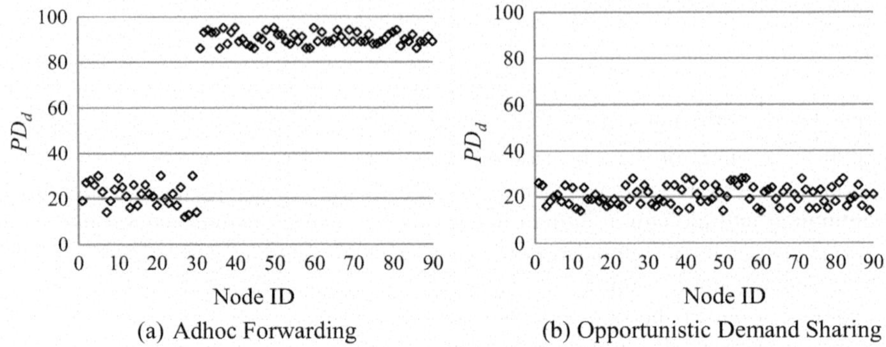

Fig. 3.5 Perception deviation about resource demands at shelter S_1 with multicopy scheme

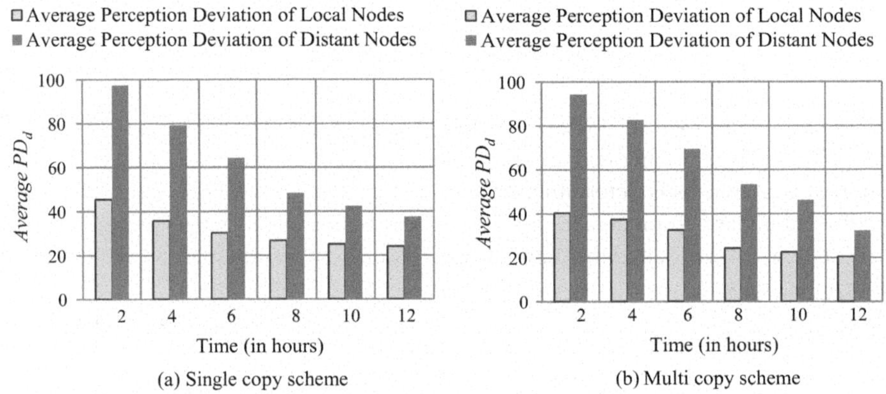

Fig. 3.6 Reduction average perception deviation with opportunistic knowledge sharing

demands at shelter S_1 against time. Figure 3.6a shows that, for single-copy scheme, the average PD_d decreases from 45 to 24% for local nodes and from 97 to 37% for distant nodes within 12 h, i.e., the operational period. Figure 3.6b shows that, for the multicopy scheme, the reduction is from 40 to 20% for local nodes and from 94 to 32% for distant nodes within 12 h. The final observation is that for both the version of the scheme, average PD_d decreases over time, reducing the perception deviation about actual resource needs of the shelters at the control station.

Average Perception Deviation Across All Shelters: Finally, the average perception deviation of all forwarder-nodes at the end of 12 h, using the single-copy opportunistic knowledge sharing scheme, is calculated across all shelters. Figure 3.7 shows that the perception deviation remains consistent from 12 to 32% for all forwarder-nodes and across all shelters. Therefore, it is evident that the proposed scheme has a maximum percentage deviation of 32% which is good enough considering the impaired transmission facilities and challenges in communication.

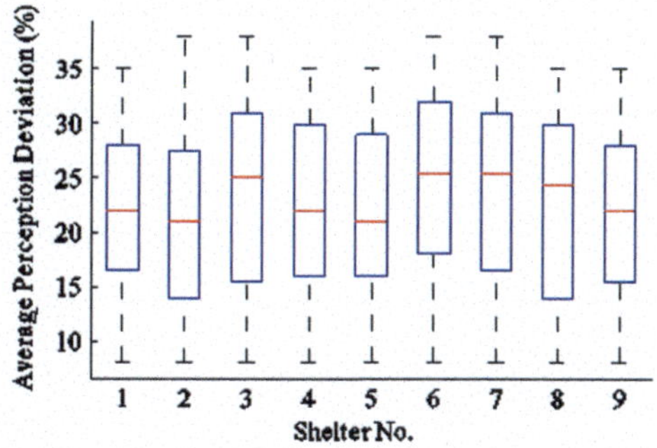

Fig. 3.7 Perception deviation for all nine shelters

3.5 Conclusion

In this chapter, assuming an underlying DTN leveraged smartphone (carried by volunteers and relief workers)-based communication network, an opportunistic knowledge sharing scheme is proposed. In this proposed scheme, when any one of the forwarder-nodes manages to visit the control station (located outside the affected area), it delivers its knowledge about the global view of resource needs therein. Here, only one volunteer's visit to the control station is sufficient for delivering a near accurate view of the unified resource needs of the entire area. Simulation results justify that the scheme succeeds to transmit resource needs to the control station even in a sparse network environment with an average error of 18.9% within the operational period. Although such need information may not be accurate to the highest level of precision but can always serve as a basis for resource allocation to a far-flung shelter whose requirements were otherwise unknown. Thus, the opportunistic knowledge sharing scheme assists in successful accumulation of emergency resource needs at the control station. The next important task of the control station is to utilize these accumulated needs judiciously, for effective resource planning.

References

1. Basu S, Roy S, DasBit S, Bandyopadhyay S (2016) A human mobility based knowledge sharing approach for post disaster need assessment using DTN. In: Proceedings of ACM ICDCN, Article 34
2. Mukherjee A, Basu S, Roy S, Bandyopadhyay S (2015) Developing a coherent global view for post disaster situation awareness using opportunistic network. Proceedings of IEEE COMSNETS 2015:1–7
3. Fall K et al (2010) A disruption-tolerant architecture for secure and efficient disaster response communications. In: Proceedings of ISCRAM
4. Paul PS et al (2015) Challenges in designing testbed for evaluating delay-tolerant hybrid networks. In: Proceedings of PerCom workshops, pp 280–283
5. Hossmann T et al (2011) Twitter in disaster mode: security architecture. In: Proceedings of SWID, Article No. 7

6. Mehendale H, Paranjpe A, Vempala S (2011) LifeNet: a flexible Ad hoc networking solution for transient environments. SIGCOMM Comput Commun Rev 41(4):446–447
7. Ntareme H, Zennaro M, Pehrson B (2011) Delay tolerant network on smartphones: applications for communication challenged areas. In: Proceedings of ExtremeCom, Article No. 14
8. Wang S et al (2013) Opportunistic Routing in Intermittently Connected Mobile P2P Networks. IEEE J Sel Areas Commun 31(9):369–378
9. Wang S, Liu M, Cheng X, Song M (2012) Routing in pocket switched networks. IEEE Wirel Commun 19(2)
10. Henderson T, Kotz, Abyzov I, Yeo J (2005) CRAWDAD Trace Set Dartmouth/campus/movement (v. 2005-03-08). Available via. https://crawdad.org/dartmouth/campus/20090909/. Accessed Jan 2015
11. Gonzalez MC, Hidalgo CA, Barabasi AL (2008) Understanding individual human mobility patterns. Nature 453(7196):779–782
12. Song C, Qu Z, Blumm N, Barabasi A (2010) Limits of predictability in human mobility. Science 327(968):1018
13. Uddin MYS, Nicol DM, Abdelzaher TF, Kravets RH (2009) A post-disaster mobility model for delay tolerant networking. In: Proceedings of winter simulation conference, pp 2785–2796
14. Nelson SC, Harris AF, Kravets R (2007) Event-driven, Role-based mobility in disaster recovery networks. In Proceedings of CHANTS 2007, pp 27–34
15. Aschenbruck N, Gerhards-Padilla E, Martini P (2009) Modeling Mobility in Disaster Area Scenarios. Perform Eval J 66(12):773–790
16. Aschenbruck N, Gerhards-Padilla E, Martini P (2008) A survey on mobility models for performance analysis in tactical mobile networks. J Telecommun Inf Technol 2(2008):54–61
17. Bai F, Helmy A (2004) A survey of mobility models. Wirel Adhoc Netw 206:1–30
18. Keränen A, Ott J, Kärkkäinen T (2009) The ONE simulator for DTN protocol evaluation. In: Proceedings of SIMUTOOLS, Article No. 55
19. Vahdat A, Becker D (2000) Epidemic routing for partially-connected Ad Hoc networks. Master's Thesis, Duke University, Durham, NC, USA
20. Lindgren A, Doria A, Schelén O (2003) Probabilistic routing in intermittently connected networks. SIGMOBILE Mob Comput Commun Rev 7(3):19–20
21. Spyropoulos T, Psounis K, Raghavendra CS (2005) Spray and wait: an efficient routing scheme for intermittently connected mobile networks. In: Proceedings of SIGCOMM workshop on delay-tolerant networking, pp 252–259
22. Burgess J, Gallagher B, Jensen D, Levine BN (2006) MaxProp: Routing for vehicle-based disruption-tolerant networks. In: Proceedings of INFOCOM, pp 1–11
23. Balasubramanian A, Levine B, Venkataramani A (2007) DTN routing as a resource allocation problem. SIGCOMM Comput Commun Rev 37(4):373–384
24. Daly EM, Haahr M (2007) Social network analysis for routing in disconnected delay-tolerant MANETs. In: Proceedings of MobiHoc, pp 32–40
25. Hui P, Crowcroft J, Yoneki E (2008) BUBBLE Rap: social-based forwarding in delay tolerant networks. In: Proceedings of MobiHoc, pp 241–250
26. Nelson SC, Bakht M, Kravets R (2009) Encounter-based routing in DTNs. In: Proceedings of INFOCOM, pp 846–854
27. Watts J, Strogatz SH (1998) Collective dynamics of 'Smallworld' networks. Nature 393(6684):440–442
28. Wei K et al (2015) Exploiting Small world properties for message forwarding in delay tolerant networks. IEEE Trans Comput 64(10):2809–2818
29. Kleinberg J (2000) The small-world phenomenon: an algorithmic perspective. In: Proceedings of STOC, pp 163–170
30. Ahmed S, Kanhere SS (2010) Characterization of a large-scale delay tolerant network. In: Proceedings of LCN, pp 56–63
31. Bhattacharjee S, Basu S, Roy S, DasBit S (2016) Best-effort delivery of emergency messages in Post-Disaster Scenario with content-based filtering and priority-enhanced PRoPHET over DTN. In: Proceedings of IEEE COMSNETS, pp 1–7

32. Bhattacharjee S, Roy S, Bandyopadhyay S (2014) Exploring an energy-efficient DTN framework supporting disaster management services in post disaster relief operation. Wirel Netw 21(3):1033–104
33. Wang W, Zwolinski M (2013) An improved instruction-level power model for ARM11 microprocessor. In: Proceedings of HIP3ES, pp 1–7
34. Google Map of Water, Food, Shelter and Medical Resources for Nepal Earthquake. (2015). Available via https://www.google.com/maps/d/viewer?mid=1Iv7GILViqyJAFn5o5hi1F2Fg8mc&hl=en_US. Accessed July 2016

Chapter 4
Resource Planning

The opportunistic knowledge sharing scheme, proposed in Chap. 3, is used to accumulate the emergency resource needs at the control station, over a smartphone-based DTN, for necessary resource planning. Due to typical characteristics of DTNs [1–3], the control station is likely to receive some corrupted or stale information. Even possibility of receiving no information cannot be ruled out. Thus, the opportunistically received resource needs are required to be thoroughly validated before these are used for resource allocation. Furthermore, resources in a post-disaster scenario are scarce and not all demands can be met. Minimizing the deficit in allocation of high-utility resources is crucial. On the other hand, since the underlying communication environment uses DTN, information about resource needs at the shelters reaches significantly late at the control station. Hence, minimizing the resource deployment time is inevitable.

In this chapter, a resource planning mechanism is presented which is performed in two phases. First, a case-based reasoning (CBR)-driven need validation scheme that is executed at the control station is proposed. Next, a utility-based integer programming model is formulated using the opportunistically transmitted and CBR-validated resource needs for optimal resource allocation. The formulated model reduces the resource deficit as a whole and the total resource deployment time to achieve fast and effective disaster relief. The control-node (a computer/server at the control station) receives current resource demands from the forwarder-nodes, carries out a CBR-driven technique to validate/estimate the needs and performs a utility-driven optimal resource allocation strategy to minimize resource deficit and resource deployment time. Optimal resource allocation through CBR-driven need validation scheme and the integer programming model bring about efficient *resource planning*.

S. Basu et al., *Reliable Post Disaster Services over Smartphone Based DTN*, Smart Innovation, Systems and Technologies 137,
https://doi.org/10.1007/978-981-13-6573-7_4

4.1 Literature Review

In this section, first, some literature on CBR-driven resource demand forecasting is reviewed. Subsequently, some relevant resource allocation models, for both emergency and regular resources, are summarized.

4.1.1 CBR-Driven Resource Demand Forecasting

CBR is the process of solving current problems by remodelling/using previously successful solutions to similar kind of problems. In recent years, CBR is obtaining importance over other AI approaches in forecasting demand for post-disaster emergency resources because it is lightweight and requires little or no direct expert knowledge acquisition. Some of the CBR-driven demand forecasting techniques for emergency resources are reviewed in this sub section.

Zhao et al. in [4] provide an overview on the recent developments in emergency resource demand forecasting methods and analysed that the forecasting accuracy is highly correlated with relief classification, the characteristics of relief-demand in the disaster-struck area and the selection of relief-demand forecasting method. Authors perform categorization of relief resources, explain the relationship between categorized relief resources and demand forecasting and finally elaborate the application of CBR and information entropy theory to provide reference for relief distribution.

In the work [5] of Liu et al., the case-based reasoning prediction method associated with risk analysis process is used to conduct demand prediction on the quantity, quality and type of emergency resources. After conducting risk analysis on target area, characteristic values of risk in that area can be obtained, including possible incident type, incident results and occurrence probability of incident. Accordingly, the case-based reasoning process can be used for emergency resource demand prediction.

Deqiang et al. in [6] present a precise prediction scheme based on CBR and back-propagation (BP) neural network. This scheme selects samples of emergency supplies which are used in BP neural network training. As similar emergency supply demand processes are bound to have similar demand law, the method first puts forward a set of base cases which are similar to the goal emergency supply demand, then finds out the demand law of base cases by using BP neural network and finally applies the law to predict the goal case of emergency demand.

Duan et al. in [7] use CBR and grey relationship analysis to predict the demand for emergency resources. CBR is used to calculate the similarity matrix between historical cases and target cases. The similarity vector is calculated next. The grey relationships between historical cases and target cases are analysed, and then the model is built where the similarity and grey relationships are integrated; the cases are verified by the model at the end.

Based on the above study, it is observed that CBR has found wide application in emergency resource demand forecasting. However, there has been no work on harnessing such forecasting capacity of CBR to formulate a need validation technique that validates, rectifies and forecasts future needs for critical resources at the control station.

4.1.2 Resource Allocation

In this subsection, selected works on resource allocation models for both emergency and non-emergency resources are reviewed. Primarily, research results on utility-based resource allocation are reviewed. Subsequently, current researches on the applications of operations research techniques in post-disaster emergency resource allocation are explored.

4.1.2.1 Utility-Based Models

Network resource allocation models based on utility of respective resources have been extensively explored by researchers in the past. Such allocation models essentially employ utility functions derived from various network parameters like transmission rates, potential delays and quality of service. Four utility-based network resource allocation models are outlined in this subsection.

Kuo and Liao in [8] study utility-based maximization for resource allocation in the downlink direction of centralized wireless networks. Authors consider two types of traffic, i.e., best effort and hard QoS, and develop some essential theorems for optimal wireless resource allocation. Three allocation schemes are then proposed. The performance of the proposed schemes is evaluated via simulations. The results show that optimal wireless resource allocation is dependent on traffic types, total available resource and channel quality, rather than solely dependent on the channel quality or traffic types as assumed in most of the existing work.

Sheikh et al. in [9] propose a link utility function based on which the resource negotiations are done. The utility function models the preferences of nodes in allocating various PHY layer resources to throughput. The utility functions are also used to quantify the performance benefits and power costs associated with the allocation prior to selecting the optimal operating point.

Song et al. in [10] discuss downlink resource allocation and scheduling for OFDM-based broadband wireless networks. Authors present a cross-layer resource management framework leveraged by utility optimization. It includes utility-based resource management, QoS architecture and resource allocation algorithms. The framework exploits wireless channel and queue information for rate-based and delay-based multichannel scheduling and includes theoretical exploration of the fundamental mechanisms in wireless resource management. They also present a solution for efficient resource allocation for heterogeneous traffic with diverse QoS requirements.

Briceno et al. in [11] propose a time-utility functions for modelling and evaluating resource allocations in a heterogeneous computing system. Authors derive a mathematical model of a heterogeneous computing system where tasks arrive dynamically and need to be assigned based on their priority, precedence, utility characteristic class and task execution type. Priority and precedence are used to generate time-utility functions that describe the value of a task at any given time. A performance metric based on the total utility gained from completing tasks is derived and used for comparing the performance of resource allocation heuristics in this environment.

Although there is substantial work done on utility-based network resource allocation, to the best of the knowledge, there has been no work on utility-based emergency resource allocation in a post-disaster environment.

4.1.2.2 Operations Research-Based Models

There has been extensive studies on application of operations research-based techniques in disaster operations management. Some works on the operations research-based emergency resource allocation models are examined in this sub section.

Balbis et al. in [12] provide a general framework for the formalization of problems relevant to emergency management through real-time resource assignment. A centralized decision model is considered, in which a single decision centre is assumed to be able to manage several kinds of emergency resources. Then, a dynamic resource management problem is formalized, whose objective is that of minimizing the effects of a natural disaster on a territory.

To improve the identification and management of response assets in a mass-casualty incident, as well as to help coordinate the initial response, Kondaveti et al. in [13] propose a decision support system for resource allocation in disaster management. Using information collected from emergency response agencies, operations research techniques are used to find an optimal solution for allocation of resources to the disaster site.

Altay in another work [14] resolves the capability-based multiresource multilocation allocation problem using an integer programming model. The model uses a nationwide resource inventory listing to allocate response personnel, facilities, supplies and equipment to impact jurisdictions. Two variants of a capability-based multiresource multilocation resource allocation model are proposed. One version of the model assumes that resources are plentiful and readily available, while the other considers the case where the demand for resources overwhelms supply.

Based on the actual demand of emergency resources, Yang et al. in [15] propose a multi-emergency dynamic allocation model. The model considers that the casualties of trapped victims change over time. This optimization model for the allocation of rescue resources aims at maximizing the overall emergency rescue effectiveness of rescue proposal of the allocated and the unallocated resources at each stage during the planning period. The model also allows for the cost of allocation resources. Considering the purchasing cost, inventory cost and opportunity loss cost during the cost objective construction, a method that combines ideal point method and unit cost utility method is used to solve the allocation model.

In another work under this category [16], Muaafa et al. focus on developing a multi-objective optimization model and an evolutionary algorithm as a first step to generate optimal emergency medical response strategies characterized by the selection of: (1) locations of temporary emergency units, (2) dispatching strategies of emergency vehicles to evacuate injured victims to the temporary emergency units and (3) number of victims to evacuate to each unit. The objectives of the optimization model are to minimize response time and cost of the response strategy. The evolutionary algorithm is used to solve the model and find a set of Pareto optimal solutions where each solution represents a different emergency medical response strategy.

The objective of the work presented by Chacko et al. in [17] is to outline an operations research-based decision support system, with a resource allocation model in its model base, wherein the model can be extended to include a multihazard scenario. The work is further focused on long-term disaster operations management (DOM) planning models under features unique to a multihazard scenario.

Although the above works successfully propose optimal resource allocation models for emergency resources, none of them explicitly calculates the exact utility of emergency resources; instead, they use subjective utility values. Also, the dynamic nature of resource needs is not considered in any of these works.

4.2 Case-Based Reasoning-Driven Need Validation Technique

Need validation can be performed only by comparing the received need with forecasted need, guessed on the basis of archived data available at the control station. Thus, need validation requires prediction of need based on historic/archived data. In recent years, case-based reasoning (CBR) is obtaining importance over other history-based prediction techniques because it is lightweight and does not require much expert knowledge acquisition. CBR therefore appears to be a better approach than other rule induction algorithms for rich, complex domains where there are numerous ways to generalize a case. These facts act as motivating factors to exploit the prediction capacities of CBR to propose a need validation technique that validates, rectifies and anticipates future needs for critical resources at the control station.

In this subsection, the CBR-driven technique [18] for validating emergency resource demands is explained. The need validation technique (a) validates the received resource demands, (b) rectifies the corrupted or stale demands and (c) projects the unavailable demands. The technique can be applied for validating resource utilities likewise. Once a current demand set reaches the control station, using the opportunistic knowledge sharing algorithm discussed in Chap. 3, the control-node uses those demands to prepare a global demand matrix. Such a matrix consists of various resource demands obtained from different shelters. It is to be noted that, a received demand set may not be complete, in the sense that, it may not contain demand for resources at all shelters in all regions. If multiple demand sets (possibly different) are received within a given interval, recency $t_{(k,i)}$, of each demand vector in the received sets is checked so that only latest demands are considered. This global demand matrix will be used for verification, rectification and projection of shelter demands.

4.2.1 Case-Based Reasoning

The core of case-based reasoning is the case base, which is a collection of previously made and stored experience items, called cases. CBR solves a new problem in four steps [19, 20]—by finding a similar past case(s), and reusing them in the new problem situation—RETRIEVE; once similar cases are identified, the appropriate solution(s) from the case base are adapted to solve the current problem—REUSE; adjust the proposed solution for further accuracy—REVISE; when a new (successful) solution to the new problem is found, it is stored in the case base to increase its competence, thus implementing a learning behaviour—RETAIN. In our case, various emergency resource demands at different shelters for previous time instances are recorded in the case base. The values of the situational parameters influencing various resource demands at these time instances are also recorded in the case base. When a new demand for an emergency resource is arrived at the control station through opportunistic demand sharing, similar cases are retrieved from the stored case base and the corresponding resource demand for the current time point is predicted based on the archived data. This predicted value is used to substantiate the new demand information. The validated demand is then recorded in the case base and can be used for future demand prediction. Figure 4.1 depicts the complete CBR cycle for demand validation.

4.2.2 Demand Validation Using CBR

Let D_{rs} be the demand for the emergency resource r in the shelter s for n time points t_1 to t_n. Demand D_{rs} is stored at the control station. It is further assumed that $\{ISP_1, ISP_2, \ldots, ISP_m\}$ are the values of m influential situational parameters ($ISPs$) stored for each of the n time points. Demand D_{rs} at time point t_i is denoted as $D_{t_i rs}$, and value of the influencing parameter ISP_j corresponding to $D_{t_i rs}$ is denoted as $ISP_{t_i j}$, i =1, 2, …, n and j =1, 2, …, m. The entire case base is denoted as $\left\{C_{t_i} : D_{t_i rs}, ISP_{t_i 1}, ISP_{t_i 2}, \ldots, ISP_{t_i m}\right\}_{i=1}^{n}$, where C_{t_i} is the case related to time point t_i. Next, a weight is attached with each of these $ISPs$ to indicate its importance in determining D_{rs}. To determine such importance, the correlation between the $ISPs$ and D_{rs} is used. Thus, weight of a ISP_j can be defined as:

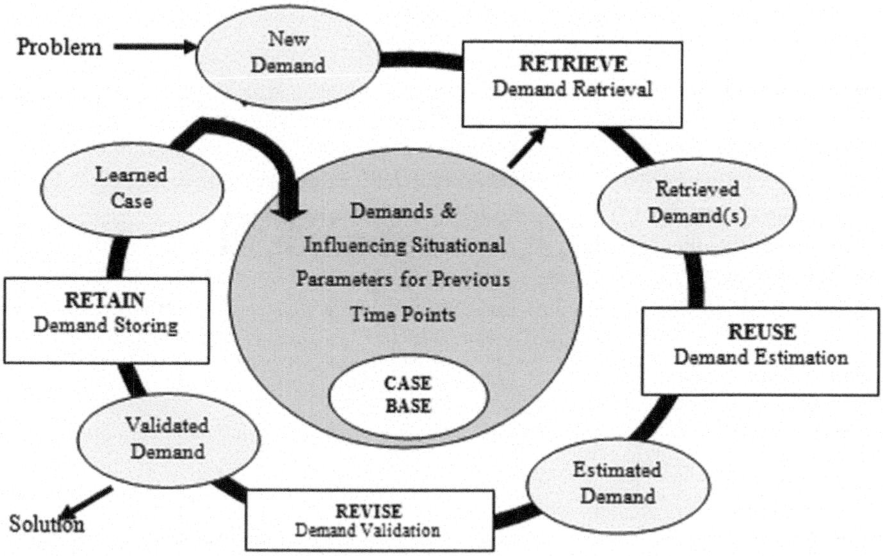

Fig. 4.1 CBR cycle for demand validation

$$w_j = \frac{\sum_{i=1}^{n} \left(ISP_{t_i j} - \overline{ISP_j}\right)\left(D_{t_i rs} - \overline{D}_{rs}\right)}{s_{F_j} \times s_{D_{rs}}} \tag{4.1}$$

where $\overline{ISP_j}$ and $\overline{D}_{rs}$ are means, s_{F_j} is the standard deviations of parameter ISP_j, and $s_{D_{rs}}$ is the standard deviations of demand D_{rs}. These weights are also stored in the case base and are used while estimating demands.

As the demand, $D_{t_{n+1} rs}^{received}$, for time point, t_{n+1}, predicted by the PCR model and communicated by the opportunistic demand sharing scheme is received by the control station, it gets verified before using it for resource planning. CBR is used for forecasting the demand at time t_{n+1} which is used for verifying the received demand. $\left\{C_{t_{n+1}} : D_{t_{n+1} rs}^{received}, ISP_{t_{n+1}1}, ISP_{t_{n+1}2}, \ldots, ISP_{t_{n+1}m}\right\}$ is then considered as a new case pertaining to time point t_{n+1}. However, the parameter values $\left\{ISP_{t_{n+1}1}, ISP_{t_{n+1}2}, \ldots, ISP_{t_{n+1}m}\right\}$ are neither available in the case base nor received from the shelters. Therefore, the next task is to estimate the values of these parameters for time point t_{n+1} based on their values at time points t_1–t_n. Owing to the delay in transmission over the DTN, these time points may not be equidistant. It is therefore reasonable to use Lagrange's interpolation formula for such extrapolation. Hence, $ISP_{t_{n+1}j}$ can be extrapolated for time point t_{n+1} as

$$\begin{aligned} ISP_{t_{n+1}j} &= \frac{(t_{n+1} - t_2)(t_{n+1} - t_3)\ldots(t_{n+1} - t_n)}{(t_1 - t_2)(t_1 - t_3)\ldots(t_1 - t_n)} ISP_{t_1 j} + \cdots \\ &+ \frac{(t_{n+1} - t_1)(t_{n+1} - t_2)\ldots(t_{n+1} - t_{n-1})}{(t_n - t_1)(t_n - t_2)\ldots(t_n - t_{n-1})} ISP_{t_n j} \end{aligned}$$

$$\text{i.e., } ISP_{t_{n+1}j} = \sum_{i=1}^{n} \prod_{\substack{l=1 \\ i \neq l}}^{n} \frac{t_{n+1} - t_l}{t_i - t_l} ISP_{t_i j}, \quad j = 1, 2, \ldots, m \tag{4.2}$$

The extrapolated $ISP_{t_{n+1}j}$ may be used as the value of ISP_j corresponding to $D_{rs}^{received}$. Such extrapolation prevents transmission of ISP values related to $D_{rs}^{received}$ to the control station over the DTN, thus reducing network overhead.

Subsequently, the new case $C_{t_{n+1}}$ is compared with the cases C_{t_i} stored in the case base to find a case from the case base that is almost similar to the new case. Corresponding parameters of new case and archived cases are compared for such similarity assessment. The similarity between two cases is computed as

$$sim\left(C_{t_{n+1}}, C_{t_i}\right) = \frac{1}{\sum_{j=1}^{m} w_j} \sqrt{\sum_{j=1}^{m} w_j \left(ISP_{t_{n+1}j} - ISP_{t_i j}\right)^2}, \quad i = 1, 2, \ldots, n \tag{4.3}$$

where $\sqrt{\sum_{j=1}^{m} w_j \left(ISP_{t_{n+1}j} - ISP_{t_i j}\right)^2}$ is the Euclidean distance of the weighted differences defined by the attributes $ISP_j s$. All the cases with a nonzero $sim\left(C_{t_{n+1}}, C_{t_i}\right)$ are considered as similar cases, among which $sim^{\max}\left(C_{t_{n+1}}, C_{t_i}\right)$ is the maximum similar case. Accordingly, the resource demand of the most similar case is taken as an estimate, $D_{t_{n+1}rs}^{estimated}$, of the emergency resource demand for case $C_{t_{n+1}}$ at time point t_{n+1}. Three different cases may arise while verifying the received demand $D_{t_{n+1}rs}^{received}$ with the estimated demand $D_{t_{n+1}rs}^{received}$. in terms of relative error (RE), where

$$RE = \frac{\left| D_{t_{n+1}rs}^{received} - D_{t_{n+1}rs}^{estimated} \right|}{D_{t_{n+1}rs}^{estimated}}$$

Each case and its corresponding decision are elaborated below:

Case I: $RE \leq \gamma$, where γ is a threshold value
Decision:
Use $D_{t_{n+1}rs}^{received}$ as the demand for the resource r at the shelter s for time point t_{n+1}. Add $\left\{C_{t_{n+1}} : D_{t_{n+1}rs}^{received}, ISP_{t_{n+1}1}, ISP_{t_{n+1}2}, \ldots, ISP_{t_{n+1}m}\right\}$ to the case base for future use.

Case II: $RE > \gamma$
Decision:
Use $D_{t_{n+1}rs}^{estimated}$ as the demand for the resource r at the shelter s for time point t_{n+1}. $D_{t_{n+1}rs}^{received}$ i.e., the demand predicted by the PCR model and communicated through opportunistic demand sharing scheme to the control station might have been corrupted during transmission. If such anomaly persists for a few subsequent time instances for a given shelter, the control station contacts the corresponding shelter to verify the reason for such inconsistency.

Case III: $D_{t_{n+1}rs}^{received}$ is absent in the global demand matrix
Decision:

Use $D^{estimated}_{t_{n+1}rs}$ as the demand for resource r at the shelter s for time point t_{n+1}. The communicated demands might have been lost due to transmission error in the network.

Thus, the CBR-driven need validation scheme may be used to verify received demands and calculate missing demands so as to get an exact and comprehensive picture of resource needs at all shelters in the disaster-struck area. Such an exhaustive depiction helps in coherent and accurate resource allocation.

4.2.3 *Performance Analysis*

The efficiency of the proposed CBR-assisted demand validation technique is evaluated through quantitative analysis in this section.

4.2.3.1 Simulation Environment and Data

The simulation environment and data used for performance analysis of the PCR-based demand forecasting model, described in Chap. 2 (Sects. 2.2.4.1 and 2.2.4.2), are used for conducting quantitative performance analysis of the CBR driven demand validation technique as well.

4.2.3.2 Simulation Set-up

For the CBR-driven need validation scheme, the average resource demands and the average values of the influencing situational parameters as depicted in Tables 2.4 and 2.5 (Sect. 2.2.4.2 of Chap. 2), respectively, are assumed to be applicable to all the shelters in Kathmandu. The demand for rice (R_1) at shelters of Kathmandu (Table 2.4) and values of ISPs (Table 2.5) are used to assess the effectiveness of CBR-driven need validation technique. Demands for weeks 8–14 are guesstimated considering the data for weeks 1–7 as case base. The weight of each parameter is computed using Eq. (4.1), and the values of the influencing situational parameters ISP_1 to ISP_{10} for weeks 8–14 are extrapolated using Eq. (4.2). Next, the similar cases corresponding to the above weeks are obtained using Eq. (4.3) to approximately calculate the corresponding demands. Similar CBR estimates are derived for resource R_1 at shelters of Nuwakot and Sindhupalchowk. The prepared case base is then used to estimate the demand for resources rice (R_1), sugar (R_2) and biscuits (R_3) at shelters of Kathmandu, Nuwakot and Sindhupalchowk for week 15. As discussed in Sect. 3.4.2.4 of Chap. 3, after 12 h of operation of the opportunistic knowledge sharing scheme, a global demand matrix gets generated at the control-node, using the opportunistically received resource demands over a smartphone-based DTN. The global demand matrix, shown in Table 4.1, contains demands for R_1, R_2, and R_3 at

Table 4.1 Global demand matrix

Resource	Demand in shelter								
	Kathmandu			Sindhupalchowk			Nuwakot		
	S_1	S_2	S_3	S_4	S_5	S_6	S_7	S_8	S_9
R_1	52	**105**	52	72	72	72	**135**	66	66
R_2	1	1	**9**	4	–	4	3	3	3
R_3	**56**	30	30	**99**	51	51	49	**123**	–

shelters of Kathmandu, Nuwakot and Sindhupalchowk. The CBR estimates for week 15 are used to validate the demands in the global demand matrix, before they are used for resource allocation.

4.2.3.3 Simulation Metrics

The three simulation metrics used for assessing the performance of the PCR-based demand forecasting model, described in Chap. 2 (Sect. 2.2.4.4), are used for conducting quantitative performance analysis of the CBR-assisted demand validation technique as well. However, the significances of the parameters used in the metrics are different in case of evaluating the effectiveness of the CBR-assisted need validation technique. The metrics are redefined as below:

– ***Percentage Bias (PB)*** $= \frac{1}{\sum_{i=1}^{n} D_{t_i rs}} \sum_{i=1}^{n} \left(D_{t_i rs} - \widehat{D}_{t_i rs}\right) \times 100$

where $D_{t_i rs}$ and $\widehat{D}_{t_i rs}$ denote observed and CBR-estimated needs, respectively. The ideal value of *PB* is 0 which indicates a perfect concurrence between the forecast and observed needs. Negative and positive values signify over- and under-estimation, respectively.

– ***Nash–Sutcliffe Efficiency (NSE)*** $= 1 - \left[\frac{\sum_{i=1}^{n}\left(D_{t_i rs} - \widehat{D}_{t_i rs}\right)^2}{\sum_{i=1}^{n}\left(D_{t_i rs} - \overline{D_{rs}}\right)^2}\right]$

where $\overline{D_{rs}}$ denotes the average of observed needs. The ideal value of NSE is 1 that indicates a perfect agreement between forecast to observed needs; a value between 0 and 1 indicates an adequate performance.

– ***Accuracy Factor (AF)*** $= 10 \times \frac{1}{n} \sum_{i=1}^{n} \left|\log \frac{\widehat{D}_{t_i rs}}{D_{t_i rs}}\right|$

The ideal value of AF is 1 that indicates perfect match between the forecast and observed values. The smaller the value of AF, the more accurate is the forecasts.

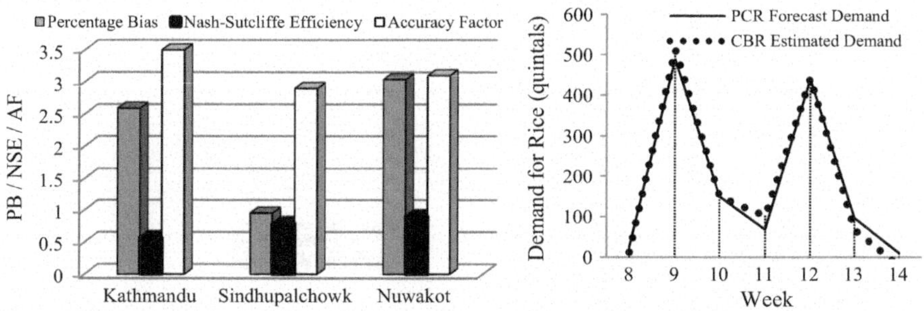

(a) PB, NSE, and AF of CBR estimated demands

(b) CBR estimated demands vs. PCR forecast demands for weeks 8-14 in shelters of Kathmandu

Fig. 4.2 Performance of CBR estimates for resource R_1

4.2.3.4 Simulation Results

In this first step of simulation, the efficacy of the CBR technique in estimating future demands is evaluated. Subsequently, the prepared case base is used to validate the demand values received through the opportunistic knowledge sharing scheme.

- ***Evaluation of the CBR Technique***: Performance of the CBR technique with respect to the observed demands of resource R_1 is plotted in Fig. 4.2a. It is found from Fig. 4.2a that the average *PB* is 2.19, average *NSE* is 0.75, and average *AF* is 3.16, all of which are within the acceptable limits as indicated in Sect. 2.2.4.4 of Chap. 2. The efficiency of the CBR technique with respect to the PCR forecast demands of resource R_1 at shelters of Kathmandu for week 15, as derived in Sect. 2.2.4.5 of Chap. 2, is plotted in Fig. 4.2b. It presents the demand for rice (R_1) in shelters of Kathmandu during weeks 8–14. Figure 4.2b indicates that the CBR-estimated demands for these weeks, and the corresponding PCR forecasts are in close concurrence, with an average relative error of 3.18%. This makes the CBR-estimated demands perfectly suitable for verifying the received PCR forecast demands.
- ***CBR-Driven Demand Validation***: Next, the CBR-assisted demand validation technique to estimate the demand for resources R_1, R_2 and R_3 at shelters of Kathmandu, Nuwakot and Sindhupalchowk for week 15. These estimated demands are compared with received demands found in the global demand matrix, shown in Table 4.1. While verification, it is found that some received demand values in the global demand matrix (highlighted in bold) are corrupted or stale and some others are missing, owing to message drop or any other network issue. Necessary actions are taken based on the decision rules explained in Sect. 4.2.2, taking the threshold value $\gamma = 1$. Table 4.2 shows the comparison results and corresponding actions. Thus, the proposed CBR-based demand validation scheme validates the

Table 4.2 CBR-driven validation decision for week 15

Resource	Demand in Shelter								
	Kathmandu			Sindhupalchowk			Nuwakot		
	S_1	S_2	S_3	S_4	S_5	S_6	S_7	S_8	S_9
R_1	52 / 50 / Received	**105** / 50 / Estimated	52 / 49 / Received	72 / 69 / Received	72 / 72 / Received	72 / 70 / Received	**135** / 65 / Estimated	66 / 66 / Received	66 / 68 / Received
R_2	1 / 3 / Received	1 / 2 / Received	**9** / 3 / Estimated	4 / 4 / Received	**Missing** / 3 / Estimated	4 / 4 / Received	3 / 4 / Received	3 / 2 / Received	3 / 5 / Received
R_3	**56** / 27 / Estimated	30 / 30 / Received	30 / 32 / Received	**99** / 49 / Estimated	51 / 51 / Received	51 / 54 / Received	49 / 49 / Received	**123** / 52 / Estimated	**Missing** / 47 / Estimated

Received Demand / CBR Estimated Demand / Decision: Demand value to be used

received demands, rectifies corrupted or stale demands and calculates the unavailable demands.

In this work, a CBR-driven need validation scheme is formulated to verify the PCR forecast demands that are opportunistically transmitted to the control station over a smartphone-based DTN, before they are used for emergency resource planning. Experimental results justify that the technique projects resource needs in the shelters with visible accuracy (accuracy factor nearly 1). The technique, executed at the control station, (a) validates the observed resource needs, (b) corrects the corrupted or stale needs and (c) estimates the unavailable needs. The proposed technique enables the control station to carry out effective resource allocation. After validating the received resource needs, the next task is to formulate an optimal resource allocation model.

4.3 Utility-Based Integer Programming Model for Optimal Resource Allocation

In this section, a utility-based integer programming model [21] for optimal resource allocation is formulated using the opportunistically transmitted and CBR-validated resource needs. The formulated model minimizes the overall resource deficit of high-utility resources at different shelters and also minimizes the total resource deployment time to achieve fast and effective disaster relief.

Since resources are scarce and not all demands can be met, minimizing deficit in allocation is crucial. On the other hand, since the underlying communication environment uses DTN, collection and transmission of the resource needs depend on opportunistic contacts between the forwarder-nodes. As a result, information about resource needs at the shelters reaches late at the control station. Therefore, there is no other way but minimizing the resource deployment time so that the whole endeavour of resource allocation doesn't fail to serve the purpose. Here, deployment time refers to the shipping time of resources from warehouses to shelters. In order to serve the above purposes, a utility-based integer programming model is formulated for optimal resource allocation that uses the opportunistically transmitted and CBR-validated resource needs. The formulated model has two goals—(a) minimizing overall emergency resource deficit and (b) minimizing overall resource deployment time to achieve fast and effective disaster relief.

Once the demand and utility values in the global demand matrix and global utility matrix are verified, rectified and projected by the CBR-driven need validation technique, these matrices are used as inputs to the proposed integer programming model for periodically allocating emergency resources to different shelters, from different warehouses in an optimized manner.

4.3.1 Parameters and Variables

The parameters and variables used in the proposed model are listed below.

– ***Parameters***

S	Number of shelters (shelters are considered as sinks)
R	Number of resource types
S	Number of resource warehouses (warehouses are considered as sources)
a_{rw}	Number of units of resource r available at warehouse w represents an element of the availability matrix
a_r	Total available number of units of resource type r ($a_r = \sum_{w=1}^{W} a_{rw}$)
d_{rs}	Demand for resource r at shelter s represents an element of the global demand matrix
u_{rs}	Utility of resource r at shelter s represents an element of the global utility matrix
t_{rsw}	Time required to deploy each unit of resource r to shelter s from warehouse w represents an element of the deployment time matrix
c_r	Capacity of each consignment of resource r, i.e., the maximum number of units of r that can be shipped in a consignment

– ***Variables***

x_{rs}	Number of units of resource r allocated to shelter s represents an element of the resource allocation matrix
y_{rsw}	Number of units of resource r allocated to shelter s from warehouse w represents an element of the resource deployment matrix

$$r = 1, 2, \ldots, R; s = 1, 2, \ldots, S; \text{and } w = 1, 2, \ldots, W.$$

4.3.2 *Minimizing Overall Resource Deficit*

Deficit of the resource r at the shelter s can be defined as the ratio of unfulfilled demand and actual demand, i.e.,

$$\left(\frac{d_{rs} - x_{rs}}{d_{rs}}\right)$$

To make the resource allocation strategy utility aware, resource deficit is redefined as a weighted deficit of the resource r at the shelter s, with utility as the weight, i.e.,

$$u_{rs}\left(\frac{d_{rs} - x_{rs}}{d_{rs}}\right)$$

where a higher value of u_{rs} magnifies the deficit and vice versa. The overall resource deficit is defined as the sum of resource deficits over all resources and all shelters as shown below:

$$\sum_{s=1}^{S}\sum_{r=1}^{R} u_{rs}\left(\frac{d_{rs} - x_{rs}}{d_{rs}}\right)$$

The quantity of each type of resource to be allocated to each individual shelter is determined in order to minimize the overall resource deficit subject to some constraints. The inputs to the problem are S, R, a_r, d_{rs} and u_{rs}, and the output is a resource allocation matrix. The problem is formulated as follows:

$$\text{Minimize} : \sum_{s=1}^{S}\sum_{r=1}^{R} u_{rs}\left(\frac{d_{rs} - x_{rs}}{d_{rs}}\right) \tag{4.4}$$

$$\text{Subject to} : \sum_{s=1}^{S} x_{rs} \leq a_r \quad \text{for all } r \text{ and } s \tag{4.4.1}$$

$$x_{rs} \leq d_{rs} \quad \text{for all } r \text{ and } s \tag{4.4.2}$$

$$x_{rs} \geq 0 \quad \text{for all } r \text{ and } s \tag{4.4.3}$$

4.3.3 Minimizing Total Resource Deployment Time

Since the deployment time for one unit of the resource r to the s shelter from the warehouse w is t_{rsw}, y_{rsw} units of the resource r can be deployed to shelter s from warehouse w in $t_{rsw} * y_{rsw}$ time unit. However, in the above definition of deployment time, an unrealistic assumption, i.e., shipping one unit of resource at a time, is made. In practice, multiple units of an emergency resource are shipped at a time in a single consignment. The number of units of resource r that can be shipped at a time is restricted by the capacity of a consignment, c_r in our case. Therefore, the time required for deploying y_{rsw} units of the resource r to the shelter s from the warehouse w is redefined as follows:

$$t_{rsw} * \left\lceil \frac{y_{rsw}}{c_r} \right\rceil$$

where $\left\lceil \frac{y_{rsw}}{c_r} \right\rceil$ is the number (nearest integer $\geq \frac{y_{rsw}}{c_r}$) of consignments necessary for shipping y_{rsw} units of the rth resource. This is justifiable as the time required for shipping 1 unit of a resource is same as what is required for shipping a consignment (containing multiple units) of that resource. In the above formulation, it is assumed that one consignment will be shipped at a time. Shipping multiple consignments at a time depends on the number of available shipment vehicles at that point of time, preparation time of the consignments and other logistic considerations which is beyond the scope of the current work. The total resource deployment time is defined as

$$\sum_{r=1}^{R} \sum_{s=1}^{S} \sum_{w=1}^{W} t_{rsw} * \left\lceil \frac{y_{rsw}}{c_r} \right\rceil$$

which is the sum total of deployment times over all resources, all shelters and all warehouses. The amount of emergency resources to be allocated to each shelter from different warehouses is determined in such a way that the total deployment time is minimized under certain constraints. The inputs to this optimization problem are S, R, W, t_{rsw}, x_{rs} and a_{rw}, and the output is a resource deployment matrix. The problem is formulated as:

$$\text{Minimize}: \sum_{r=1}^{R} \sum_{s=1}^{S} \sum_{w=1}^{W} t_{rsw} * \left\lceil \frac{y_{rsw}}{c_r} \right\rceil \tag{4.5}$$

$$\text{Subject to}: \sum_{w=1}^{W} y_{rsw} = x_{rs} \quad \text{for all } r \text{ and } s \tag{4.5.1}$$

$$\sum_{s=1}^{S} y_{rsw} \leq a_{rw} \quad \text{for all } r \text{ and } s \tag{4.5.2}$$

4.3.4 Solution of the Integer Programming Models

Integer programming is used to solve the above utility-driven optimal resource allocation model. An integer programming model is a mathematical optimization or feasibility program in which some or all of the variables are restricted to be integers. The main reason for using integer programming in this case is that the parameters and the decision variables used in the model are all integers. The above objective functions are solved using branch-and-bound methodology. Solving (4) with constraints (4.1), (4.2) and (4.3) yields a resource allocation matrix containing an optimal number (x_{rs}) of the rth resource that can be allocated to the shelter s for minimizing the overall resource deficit. Solving (5) with constraints (5.1) and (5.2) yields a resource deployment matrix containing the optimal number (y_{rsw}) of the rth resource that can be allocated to the shelter s from the warehouse w for minimizing the total deployment time.

Thus, the utility-based integer programming model for optimal resource allocation issues an optimal quantity of high-utility resources to the shelters from the nearest warehouses minimizing the overall resource deficit and the total resource deployment time.

4.3.5 Performance Analysis

The efficiency of the proposed utility-based integer programming model is evaluated through quantitative analysis in this section.

4.3.5.1 Simulation Environment and Data

The simulation environment and data used for performance analysis of the PCR-based demand forecasting model, described in Chap. 2 (Sects. 2.2.4.1 and 2.2.4.2), are used for conducting quantitative performance analysis of the utility-based integer programming model as well.

Table 4.3 (a) CBR -validated demand matrix. (b) CBR-validated utility matrix

Resource	Demand in shelter		
	Kathmandu	Sindhupalchowk	Nuwakot
	S_1	S_4	S_7
(a)			
R_1	52	72	65
R_2	1	4	3
R_3	27	49	49
(b)			
R_1	10	10	8
R_2	8	7	6
R_3	6	4	3

4.3.5.2 Simulation Set-up

The objective functions derived in (4) and (5) are solved using the LINGO [22] optimization modeling tool for integer programming. As discussed in the previous section (Sect. 4.2.3.4), the CBR-driven need validation technique produces validated global demand and utility matrices that contain accurate demands and utilities of resources rice (R_1), sugar (R_2) and biscuits (R_3) at shelters S_1 to S_9. Out of these 9 shelters, three shelters, one in Kathmandu (S_1), the other in Sindhupalchowk (S_4) and another in Nuwakot (S_7) are considered for evaluating the proposed utility-based integer programming model. The CBR-validated demand and utility matrices showing demands and utilities of R_1, R_2 and R_3 at shelters S_1, S_4 and S_7 are shown in Tables 4.3a and 4.3b, respectively. Three warehouses at Dhading (W_1), Rasuwa (W_2) and Bhakatpur (W_3) are considered. The availability matrix, shown in Table 4.4a, provides availability of different resources at the three warehouses. Table 4.4b shows the deployment time matrix that contains approximate time required for shipping different types of resources to shelters from different warehouses, depending on distances between shelters and warehouses. These tables are formed using data derived from [23–26]. The CBR -validated demand and utility matrices as well as the availability matrix are fed to LINGO for solving the objective function (4) to get the resource allocation matrix, shown in Table 4.5a. Next, the availability matrix, deployment time matrix and resource allocation matrix are fed to LINGO to get the resource deployment matrix, shown in Table 4.5b.

4.3.5.3 Simulation Metrics

Two simulation metrics, as in [13], are used to evaluate the effectiveness of the utility-based integer programming model. The metrics are defined as below:

Table 4.4 (a) Availability matrix. (b) Deployment time matrix

(a)

Shelter	Available amount of resource		
	R_1	R_2	R_3
W_1	550	360	150
W_2	450	400	180
W_3	860	240	120

(b)

Shelter	Warehouse	Deployment time of resources		
		R_1	R_2	R_3
S_1	W_1	2	6	5
	W_2	5	2	3
	W_3	3	3	5
S_4	W_1	3	5	3
	W_2	5	6	1
	W_3	1	5	3
S_7	W_1	5	1	5
	W_2	6	4	5
	W_3	2	1	4

Table 4.5 (a) Resource allocation matrix. (b) Resource deployment matrix

(a)

Shelter	Allocated amount of resource to each shelter		
	R_1	R_2	R_3
S_1	365	371	205
S_2	154	85	109
S_3	391	94	86

(b)

Shelter	Warehouse	Deployed amount of resource from each warehouse		
		R_1	R_2	R_3
S_1	W_1	50	0	16
	W_2	0	131	180
	W_3	35	240	9
S_2	W_1	0	85	134
	W_2	0	0	0
	W_3	154	0	0
S_3	W_1	0	94	0
	W_2	0	0	0
	W_3	391	0	111

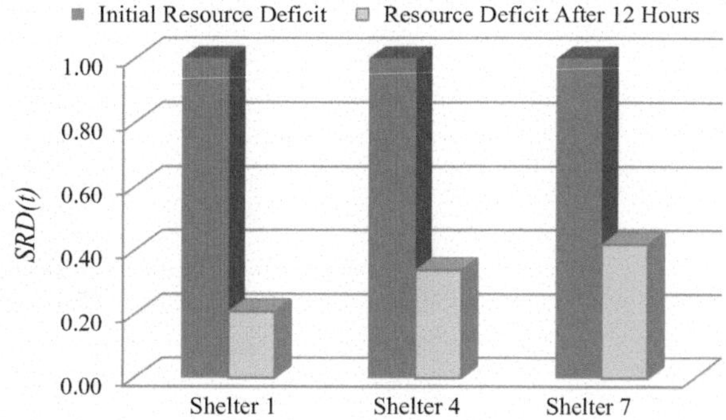

Fig. 4.3 Reduction of emergency resource deficit with time

– ***Shelter-Specific Resource Deficit* (*SRD*$_t$)** $= \frac{1}{\sum_{r=1}^{R} u_{rs}} \left(\sum_{r=1}^{R} u_{rs} * \left(1 - \frac{\sum_{w=1}^{W} y_{rs} * f_{rsw}(t)}{d_{rs}} \right) \right)$

Used to evaluate the pace at which resource deficits are getting minimized at each shelter, where $f_{rsw}(t)$ is a step function with $f_{rsw}(t) = 1$, if $t > t_{rsw}$ and $= 0$ otherwise. An ideal value of this metric is 0.

– ***Average Resource Deficit* (*ARD*$_t$)** $= \frac{\sum_{s=1}^{S} SRD_t}{S}$

Used to evaluate the rate at which overall resource deficit is getting minimized across all shelters. An ideal value of this metric also is 0.

4.3.5.4 Simulation Results

Two sets of performance evaluations are performed to estimate the effectiveness of the utility-driven integer programming model using the resource deployment matrix (Table 4.5). The first measures the rates at which deficit of emergency resources is reduced at each shelter, using the proposed model. In the next, the proposed model is compared with three other competing resource allocation models.

- ***Shelter-Specific Resource Deficit over Time***: In this experiment, SRD_t is computed after 12 h (i.e., the full operational period, as described in Sect. 3.4.1 of Chap. 3) for different time points, for the proposed system. It is observed from Fig. 4.3 that deficits in emergency resources are reduced at a higher rate (from 100 to 19%) in shelter S_1 than in shelters S_4 (from 100 to 33%) and S_7 (from 100 to 40%). This is due to the fact that utilities of resources were higher in S_1 than in shelters S_4 and S_7. Deficits at shelter S_7 are reduced at a much slower rate due to lower utility of resources therein. Thus, our proposed utility-based integer programming model for optimal resource allocation minimizes deficit of resources in those shelters first where they have higher utility.
- ***Comparative Performance***: Among the resource allocation strategies discussed in Sect. 4.1.2, the models proposed by Kondaveti et al. [13] that considers priority of resources, Altay [14] that uses capability of resources and Yang et al. [15] that uses of utility of resources are close to the proposed utility-based resource

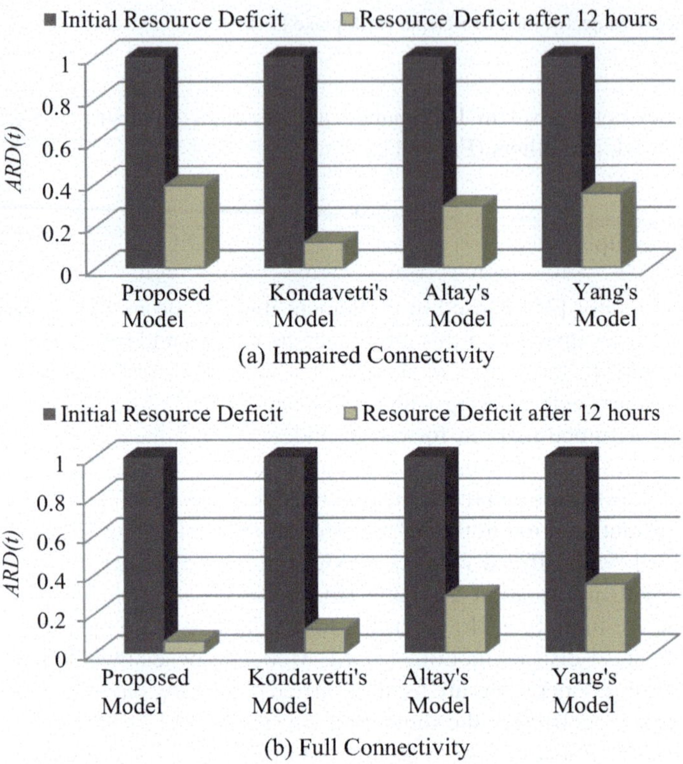

Fig. 4.4 Comparative performance of the proposed resource allocation system

allocation system. The proposed model is compared with these competing models considering two scenarios, (i) assuming a DTN environment and (ii) assuming a fully connected network. ARD_t; i.e., average resource deficit across all shelters is computed and compared for all the competing schemes including the proposed one.

Under Impaired Connectivity

In Fig. 4.4a, it is found that the performance of Kondavetti et al.'s model is best as indicated through the reduction of resource deficit by 88% and Yang et al.'s model reduces the deficit by 65%. The proposed model reduces resource deficits by only 61%. This owes to the fact that the competing models use absolutely accurate and recent needs of the shelters while allocating resources, whereas in the current model, resource allocation is based on near accurate and opportunistically received demands communicated over a DTN.

Under Full Connectivity

Configuration of parameters in ONE simulator (as described in Sect. 3.4.2.2 of Chap. 3) is modified to replicate a fully connected environment. It is observed in

Fig. 4.4b that the proposed model shows best performance by reducing resource deficits by 94%, whereas Kondavetti et al.'s model reduces deficits by 88%. Reasons are—firstly, this time the proposed model considers accurate and absolutely recent resource needs; secondly, the utility values given to the resources are realistic, depicting the actual importance of these resources at the shelters. It is, therefore, evident that given similar network conditions, the proposed model performs better than other competing models.

In this work, utility-based integer programming model for optimal resource allocation is proposed to minimize overall resource deficit and total resource deployment time. The proposed model reduces deficit of emergency resources at a much faster rate in shelters that have higher utility for them, than in shelters that have comparatively lower utility. The novelty of the model lies in the fact that it is utility driven and works even under impaired network connectivity. Simulation results justify that proposed model reduces up to 81% deficit of high-utility resources at different shelters. Moreover, the model outperforms other competing schemes by 6% in a fully connected network environment and performs almost as good as other schemes (lacks behind by only 4%) even in a sparse network environment.

4.4 Conclusion

In this chapter, first a CBR-driven need validation technique is proposed to validate the opportunistically transmitted resource needs of the shelters. Next, a utility-based integer programming model for optimal resource allocation is formulated using the CBR-validated resource needs. The CBR-driven need validation technique and the integer programming model for optimal resource allocation assist the control station in efficient resource planning to achieve fast and effective disaster relief.

It is a natural and logical inference that successful resource planning can happen only if reliable situational information, in terms of shelter resource needs, is available at the control station. However, the underlying DTN-based post-disaster communication network is highly vulnerable to several security attacks. This jeopardizes the reliability of the resource needs transmitted over the network leading to delayed and incorrect resource planning. Therefore, DTN-specific security protocols need to be designed to combat various attacks on the post-disaster communication network so that effective resource planning services can be rendered.

References

1. Venkataraman V, Lam S, Acharya HB, Shah H (2011) Delay tolerant networking—a tutorial. Available via www.cs.utexas.edu/~vinodv/files/dtn-tutorial.pdf. Accessed Jan 2017
2. Fall K, Farell S (2008) DTN: an architectural retrospective. IEEE J Sel Areas Commun 26(5):828–836

3. Warthman F (2012) Delay tolerant networks (DTNs): a tutorial. Available via http://ipnsig.org/wp-content/uploads/2012/07/DTN_Tutorial_v2.04.pdf. Accessed Jan 2017
4. Zhao J, Cao C (2015) Review of relief demand forecasting problem in emergency logistic system. J Serv Sci Manage 8(1):92–98
5. Liu W, Hu G, Li J (2012) Emergency resources demand prediction using case-based reasoning. Saf Sci 50(3):530–534
6. Deqiang F, Yun L, Changbing L (2011) Forecasting the demand for emergency supplies: based on the CBR theory and BP neural network. In: Proceedings of international conference on innovation & management, Article 45, 2011, pp 700–704
7. Duan ZP, Qian XM, Duo YQ, Wu ZZ (2014) Research on demand forecast for emergency resources based on CBR and grey relational analysis. Adv Mater Res 962–965:3050–3053
8. Kuo WH, Liao W (2007) Utility-based resource allocation in wireless networks. IEEE Trans Wirel Commun 6(10):3600–3606
9. Sheikh OM, Mahmoud SA (2012) Cross-layer design for smart routing in wireless sensor networks. Wirel Sens Networks Technol Protoc 189–214
10. Song G, Li Y (2005) Utility-based resource allocation and scheduling in OFDM-based wireless broadband networks. IEEE Commun Mag 43(12):127–134
11. Briceno LD et al (2011) Time utility functions for modeling and evaluating resource allocations in a heterogeneous computing system. In: Proceedings of IPDPSW 2011
12. Balbis L et al (2001) A decisional model for dynamic allocation of resources in natural disasters management. In: Proceedings of computer simulation in risk analysis and hazard mitigation 2001, pp 243–252
13. Kondaveti R, Ganz A (2009) Decision support system for resource allocation in disaster management. Proc EMBC 2009:3425–3428
14. Altay N (2012) Capability-based resource allocation for effective disaster response. IMA J Manage Math 24(2):253–266
15. Yang Z, Zhou H, Gao X, Liu S (2013) Multiobjective model for emergency resources allocation. Math Probl Eng 2013(2013), Article no. 538695
16. Muaafa M, Concho AL, Marquez JER (2014) Emergency resources allocation for disaster response: an evolutionary approach. Proc PSAM 2014:1–10
17. Chacko J, Rees LP, Zobel CW (2014) Improving resource allocation for disaster operations management in a multi-hazard context. Proc ISCRAM 2014:85–89
18. Basu S, Roy S, DasBit S (2018) A post disaster demand forecasting system using principal component regression analysis and case-based reasoning over smartphone based DTN. IEEE Trans Eng Manage. https://doi.org/10.1109/tem.2018.2794146
19. Benfield A (2015) Nepal earthquake event recap report. Available via http://thoughtleadership.aonbenfield.com/ documents/ 201509-nepal-earthquake.pdf. 2015. Accessed July 2016
20. Google map of water, food, shelter and medical resources for nepal earthquake. Available via https://www.google.com/maps/d/viewer?mid=1Iv7GILViqyJAFn5o5hi1F2Fg8mc&hl=en_US. 2015. Accessed July 2016
21. Nepal disaster risk reduction portal, government of nepal. Available via http://drrportal.gov.np/distributed_country. 2015. Accessed July 2016
22. Nepal earthquake recovery monitoring assessment. Shelter Cluster Nepal. Available via https://www.sheltercluster.org/sites/default/files/docs/reach_npl_report_shelter_recovery_monitoring_assessment_nov2015.pdf. 2015. Accessed July 2016
23. Basu S, Roy S, Bandyopadhyay S, DasBit S (2018) A utility driven post disaster emergency resource allocation system using DTN. IEEE Trans Syst Man Cybern Syst. https://doi.org/10.1109/tsmc.2018.2813008
24. Aamodt A, Plaza E (1994) Case-based reasoning: foundational issues, methodological variations, and system approaches. AI Commun (IOS Press) 7(1):39–59.
25. Bergmann R et al (2009) Case-based reasoning—introduction and recent developments. Künstliche Intell (Special Issue on Case-Based Reasoning) 23(1):5–11
26. LINGO - Optimization modeling software for linear, nonlinear, and integer programming, LINDO systems Inc. Available via www.lindo.com. Accessed May 2015

Chapter 5
Reliable Data Dissemination over DTN

The CBR-driven need validation technique and the utility-based integer programming model for optimal resource allocation, elaborated in Chap. 4, bring about effective resource planning. However, one of the important requirements for effective resource planning is the availability of reliable situational data, in terms of resource needs at the control station. As the resource needs are transmitted using the opportunistic knowledge sharing scheme over a DTN, reliability of the transmitted situational data depends upon the level of co-operation among participating nodes. Thus, DTNs are vulnerable to a number of security threats arising out of selfishness as well as malicious intentions of the nodes [1–3]. This leads to delayed and incorrect resource planning. Security mechanisms need to be present to combat various attacks on the post-disaster communication network so that effective resource planning services can be rendered. However, as discussed in Sect. 1.3.3 of Chap. 1, there are certain unique security challenges [4] that distinguish DTN security from conventional network security. These challenges call for a set of new security protocols to resolve security issues on the underlying DTN-leveraged post-disaster communication network from a new perspective, which is totally different from conventional network security.

In this chapter, three decentralized, lightweight and robust security solutions are proposed to suit the above requirements. First, a multilayer encryption and hashing-based scheme is proposed to protect the transmitted emergency resource needs against authenticity, eavesdropping and modification attacks. Next, a security-aided and group encounter-based PRoPHET routing protocol is proposed, where security features are incorporated into PRoPHET to provide security against eavesdropping, black hole, identity spoofing and bundle store overflow attacks. Finally, a trust-based Watchdog technique is proposed and intertwined to PRoPHET to provide security against packet drop, bad-mouthing and ballot-stuffing attacks. These security solutions run on the smartphones, carried by volunteers, and contribute towards *reliable data dissemination* so that resource planning services can be rendered reliably.

S. Basu et al., *Reliable Post Disaster Services over Smartphone Based DTN*, Smart Innovation, Systems and Technologies 137,
https://doi.org/10.1007/978-981-13-6573-7_5

5.1 Literature Review

The research on DTN security is far more challenging than conventional mobile ad hoc networks due to its unique security characteristics, as discussed in Sect. 1.3.3 of Chap. 1. These unique characteristics including long round-trip delay, frequent network partitioning, fragmentation as well as opportunistic routing make the existing security protocols designed for the conventional ad hoc networks unsuitable for DTNs. Hence, a series of new security protocols need to be carefully designed to address various security attacks [4]. Selected works on DTN security are reviewed in this section. These works are classified into two categories depending on the approaches used for attack mitigation, namely cryptographic approaches and trust-based approaches. Some works on secured keyword search are also reported.

5.1.1 Cryptographic Strategies

This section touches upon some works on DTN security that employs cryptographic strategies like encryption, hashing, signatures and public key certificates to meet the security requirements like authentication, confidentiality, integrity, anonymity and availability.

Durst, in [5], proposes a security model for DTN, based on public key technology. Essentially, a prospective user (node, process, whatever) wishing to use the services of the DTN must register its public key with a certificate authority and receive a copy of that public key signed by that certificate authority and a copy of its infrastructure credentials signed by the certificate authority. Note that this single sentence combines elements of Pretty Good Privacy (PGP) systems, public key infrastructure (PKI) systems and Kerberos systems. The user generates his own public/private key pairs, as in PGP. The certificate authority (or registration authority) verifies the identity of the prospective user and signs public keys, as in PKI.

Asokan et al. in [6] consider the applicability of identity-based cryptography (IBC) as a solution to DTN security. They investigate how security in DTNs can be bootstrapped from existing large-scale security infrastructures, like the cellular communication security infrastructure, and present an improved scheme for authentication of fragments. Authors conclude that IBC does lend to a more efficient solution to end-to-end confidentiality, in terms of server load and network connectivity requirements for recipients.

Social contacts are used for message confidentiality in delay tolerant networks in [7] by Karim El Defrawy et al. Authors aim at solving the issue where a sender tries to transmit a confidential message to a receiver through a DTN, where the sender does not share a secret key with the receiver or know the receiver's public key. In this case, the sender leverages social information, such as workplace affiliation or common social contacts, to send a confidential message to the receiver. Sender routes the confidential message using secrets shared with affiliated intermediate nodes to ultimately deliver the message to the receiver.

An onion routing technique is presented by Goldschlag et al. in [8], for anonymous communication over a computer network where messages are repeatedly encrypted and then sent through several network nodes called onion routers. Each onion router removes a layer of encryption to uncover routing instructions and sends the message to the next router where this is repeated. This prevents these intermediary nodes from knowing contents of the message.

Shikfa et al. in [9] propose an onion routing-based multiple layer commutative encryption (MLCE) scheme with the objective of providing privacy in content-based opportunistic networks. In this scheme, nodes share keys and have a regional view of the network related to a small number of hops.

The goal of the work proposed by Sakai et al. in [10] is to build performance and security models for anonymous communications in DTNs. To this end, first an abstract anonymous routing protocol based on onion routing is designed. The simplified protocol understands the performance and security issues of anonymous DTN routing, and in addition, it can be easily extended to auxiliary protocols. Then, a simplified mathematical model is proposed, which can be used to measure the performance and security guarantees of the onion-based anonymous routing in DTNs.

Edelman et al. propose a secure group communication in [11] that provides a security solution to DTN where group-based keys are exploited at the application layer and included into a larger system encrypting every message exchange. Since resources are especially limited, the implementation focuses on minimizing the traffic on the overlay associated with the maintenance of the solution.

Vakde et al. in [12] propose Privacy-Enhanced Opportunistic Networking (PEON), a system for using opportunistic networking in privacy-preserving way. PEON uses concepts of anonymous communications by routing messages through groups of peer nodes to hide the relation between the sources and the destinations. By modifying group size, the system can trade off between privacy and communication overhead. Further, individual nodes can make a similar trade-off by changing the number of intermediate groups. Authors also describe the cryptographic tools needed to facilitate changes in group membership.

Gupta et al. in [13] use AES algorithm for decentralized DTNs to secure information, where diverse key strengths deal with the security challenges in a decentralized DTN. Flexibility is given by AES to encryption and disentangling.

Cabaniss et al. in [14] propose a multiparty encryption scheme that introduces two security overlay networks capable of delivering messages securely, preventing both eavesdropping and alteration of messages. The first algorithm, chaining, uses multiple mid-points to re-encrypt the message for the destination node. The second, fragmenting, separates the message key into pieces that are routed and secured independently from each other. Both techniques improve security in hostile environments, under test conditions.

All the above researches assume a centralized security system in place for implementing different security services. Nevertheless, in a post- disaster situation, such infrastructure either rarely or may not exist. Furthermore, some of these works have high overheads in terms of computation and communication, which makes them unsuitable for the resource-limited smartphones.

5.1.2 Trust-Based Strategies

This section review works on security that employs trust-based strategies for misbehaviour detection in DTNs.

Li et al. in [1] present a trust-based scheme that can be easily put together with several existing single-copy routing protocols in opportunistic networks. The scheme targets an all-inclusive evaluation of an encounter's capability of delivering data. The scheme efficiently counters black hole and arbitrary forwarding attacks. Authors propose a positive feedback message (PFM) as the evidence of the forwarding behaviour of a node, which is fed into the Watchdog component in the framework.

Zhu et al. in [15] present iTrust, a probabilistic misbehaviour detection method, for securing routing in a DTN environment for effective trust establishment. The fundamental thought of iTrust is designing a periodically available trusted authority (TA) to review the node's activities depending on the accumulated routing proofs. iTrust is modelled as an inspection game using game theoretical analysis to show that by setting a suitable investigation likelihood, TA could secure DTN routing at a lower cost.

A trust-based cooperative Watchdog system is presented for vehicular DTNs by Dias et al. in [2] to identify and banish mischievous nodes to lessen their influence in network operations. Its functioning depends on a cooperative exchange of reputation of nodes. Identifying selfish or mischievous nodes makes it possible to improve the general network performance.

Chen et al. in [3] and [16] suggest a trust management protocol for DTNs and employ it to secure routing to show its utility. The protocol merges QoS trust with social trust to get a combined trust metric. Results attained at the design time assist dynamic trust management for DTN routing in reply to dynamically changing circumstances at run-time.

Dini et al. in [17] put forward a reputation-based protocol for different black holes. Every node locally preserves the reputation of forwarding nodes encounters, and then the node chooses among those with highest reputation. The presented reputation protocol is made up of three basic methods—acknowledgments, node lists and ageing. These methods make the communication more efficient and competent of adapting to the changing operating conditions of a DTN.

Ayday et al. in [18] establish a strong and efficient security mechanism for DTNs. This consists of an iterative trust and reputation mechanism (ITRM). The mechanism enables each node to decide the trustworthiness of the nodes that it had a direct transaction. On the other hand, ITRM exploits the iterative mechanism to identify and segregate the malicious nodes from the network in a short-time.

Finally, Jethawa et al. bring in an incentive mechanism for data-centric message delivery in DTNs [19] that guarantees avoidance of congestion due to uncooperative or selfish conduct of nodes.

All the above works work out the trust/reputation of a node depending on instantaneous feedbacks given by a group of local nodes in the network. Even though such trust values may produce a local perception about the forwarding behaviour of a node, they do not portray a global view about the forwarding behaviour of the node. Furthermore, none of the schemes discussed above aim to spell out the time required for the trust values to get percolated in the network and assess the timeliness of the proposed trust protocol. The works that put together a trust framework with routing protocols like PRoPHET do not solve the self-trusting problem [1], a property of trust-based data forwarding in opportunistic networks. Self-trusting decreases delivery ratio and increases the delivery delay.

5.1.3 Secured Keyword Search

In this section, we summarize some related work on secured keyword search mainly based on a technique called public encryption with keyword search (PEKS).

Boneh et al. in [20] propose a mechanism for searching on data that is encrypted using a public key system. Authors refer to this mechanism as public key encryption with keyword search. The construction of this scheme was inspired by identity-based encryption (IBE) and aims to allow a user to search for encrypted keywords without decrypting them. In the PEKS scheme, a sender, for example, uploads an encrypted email to a server along with an encrypted list of keywords. The receiver sends the desired keyword (denoted as a trapdoor) to the email server, which then tests the encrypted emails for the presence of this trapdoor.

Hsu et al. in [21] study six existing security models of PEKS and analyse their efficiency and performance. Moreover, they conclude by deriving five security requirements that must satisfy as constructing PEKS/SCF-PEKS scheme and discuss extended issues about a keyword search scheme. Security and efficiency of the security models are also discussed.

Shikfa et al. in [22] use PEKS to allow intermediate nodes to discover partial matches between their profile and the destination profile, and use policy-based encryption to enforce confidentiality of the payload. This scheme suits for opportunistic networks well, because it has a low storage and computation overhead and it relies on an offline trusted third party (TTP) only.

Abdallah et al. in [23] aim at identifying and filling gaps with regard to consistency for PEKS. They provide a new scheme that is statistically consistent. Authors also provide a transform of an anonymous IBE scheme to a secure PEKS scheme that, unlike the previous one, guarantees consistency. Finally, three extensions of the basic notions considered here, namely anonymous HIBE, public key encryption with temporary keyword search and identity-based encryption with keyword search, are suggested for guaranteeing consistency.

Waters et al. [24] present a design for an encrypted audit log that allows a designated trusted party, the audit escrow agent, to construct keyword search capabilities, which allow (less trusted) investigators in possession of such capabilities to search for and decrypt entries matching a given keyword. The escrow agent can distribute a capability to an appropriate investigator. Since it is expected that keyword search capabilities are distributed rather infrequently, the escrow agent can be made to be very secure from attack.

Golle et al. in [25] define a security model for conjunctive keyword search over encrypted data and present a scheme for conducting such searches securely. Authors propose a first scheme for which the communication cost is linear in the number of documents, but that cost can be incurred "offline" before the conjunctive query is asked. The security of this scheme relies on the decisional Diffie–Hellman (DDH) assumption. A second scheme is proposed whose communication cost is in the order of the number of keyword fields and whose security relies on a new hardness assumption.

Once again, almost all the above works on secured keyword search use centralized security infrastructure that may not be available in a post-disaster scenario. Therefore, a decentralized PEKS-based secured message categorization technique is required that would enable forwarder-nodes to categorize situational messages, without compromising on their confidentiality, for expediting the delivery of categorized messages to their appropriate destinations.

5.2 Multilayer Encryption and Hashing-Based Scheme

There are a number of security issues that may create threats to the DTN-leveraged post-disaster communication network. One of the attacks is the authenticity attack, where a node that do not belong to any authorized rescue group may fake as a registered member of an authorized organization and claim authenticity. Hence, measuring the authenticity of a node is essential before including it in any forwarding action. Two other attacks are eavesdropping and modification attacks, where malicious nodes may eavesdrop the forwarded resource needs and maliciously alter them. Thus, confidentiality and integrity of the forwarded resource needs are to be protected until it reaches the control station.

In this section, a multilayer encryption and hashing-based scheme [26] is proposed to protect the transmitted resource needs against privacy attacks like authenticity, eavesdropping and modification. The scheme is designed to work on top of the opportunistic knowledge sharing scheme [27], proposed in Chap. 3, for transmitting and accumulating the shelter resource needs at the control station using a smartphone-based DTN. In the opportunistic knowledge sharing scheme, forwarder-nodes, i.e. volunteers carrying smartphones, maintain demand (and utility) vectors to store the demands (and utilities) of resources of its own shelter. These vectors are exchanged among local nodes, working in the same region, to create demand (and utility) matrices. These matrices are exchanged among remote nodes, working in different regions, to create demand (and utility) sets. When one of the forwarder-nodes manages to visit the control station (located outside the affected area), the node delivers its demand (and utility) set to inform the control station about the resource needs of all shelters in the entire affected area.

In the present work, first a group-based distributed authentication scheme is proposed to mitigate authenticity attack. The scheme helps forwarder-nodes to authenticate each other as registered members of legitimate rescue groups without using any central trusted authority. Next, a multilayer hashed encryption scheme, based on multilayer encryption and hashing, is proposed to mitigate eavesdropping and modification attacks. In this scheme, groups collectively work towards preserving the confidentiality and integrity of transmitted messages (i.e. resource demands and utilities). This scheme assists a node in identifying the integrity of a message at each hop without getting access to its content and taking decision on whether to forward the message to the next-hop. This makes possible early detection of corrupt messages and their elimination from the network without wasting further network resources. The schemes provide authentication, confidentiality and integrity in a fully decentralized way to go with the requirements of a DTN-based post-disaster communication network.

5.2.1 System Model

In this section, the system model related to proposed schemes is described. The system model consists of network architecture, attack model and pin distribution at the set-up phase.

5.2.1.1 Network Architecture

The underlying network architecture of the proposed multilayer encryption and hashing-based scheme is mostly the same as the network architecture used for the opportunistic knowledge sharing scheme [27] explained in Sect. 3.2.1 of Chap. 3. A typical post-disaster communication network built over a DTN, assumed for this work, consists of three categories of nodes—shelter-nodes, forwarder-nodes and control-node [27]. A post-disaster communication network, composed of shelter-nodes, forwarder-nodes and control-node, is illustrated in Fig. 3.1. The roles of these nodes are summarized below.

- ***Shelter-node***: A *shelter-node* is a laptop or workstation in the shelter. Situational parameters of a shelter like number of victims, the type and degree of emergency they have, their consumption pattern, rate of resource replenishment and rate of migration are uploaded to the shelter-node, installed therein, by the emergency managers present on site. The shelter-node dynamically forecasts the demand for different emergency resources using the proposed PCR model and enumerates the utility of such resources using the derived utility function. The shelter-node in a shelter periodically broadcasts these resource needs (i.e. demands and utilities) in the form of messages towards the forwarder-nodes that are working in and around that shelter.

- ***Forwarder-node***: *Forwarder-nodes* are the volunteers carrying DTN-enabled smartphones. These nodes work around a shelter and receive the messages, containing emergency resource needs of that shelter, broadcast by the shelter-node therein. The demands are stored in demand vectors, as shown in Table 3.1a, and utilities are stored in utility vectors, as shown in Table 3.1b. These nodes opportunistically interact with other local and remote nodes due to their mobility; automatically exchange the demand and utility vectors; build a global view of need information, in terms of demand and utility sets; and eventually transmit these sets to the control station.
  Whenever these nodes move into the radio range of each other, communication links are established. Message exchange between the DTN-enabled smartphones carried by the volunteers is accomplished by using their Bluetooth or Wi-Fi Direct interfaces, whenever two such devices come within the radio range of each other. Messages are forwarded in multiple hops on a peer-to-peer fashion using DTN routing protocols. Such Bluetooth or Wi-Fi Direct-based transmission eliminates the necessity of traditional communication requirements like GSM, GPRS or Wi-Fi. The forwarder-nodes are an essential part of any DTN as routing can be implemented only with help of these nodes. All such nodes are assumed to be resource limited in terms of processor and battery, and every such node forwards its own messages along with messages of other nodes.
- ***Control-node***: The *control-node* (a workstation or a server at the control station) receives updated resource needs, in the form of demand and utility sets, from the forwarder-nodes; carries out a case-based reasoning-driven technique to validate/estimate the needs; and executes a utility-driven optimal resource allocation strategy to allocate resources to shelters targeting minimization of resource deficit and deployment time.

5.2.1.2 Attack Model

Although there can be several types of attacks on a DTN-based post-disaster communication network, the attack model, considered in this work, consists of the following three privacy attacks:

- ***Authenticity Attack***: This is an attack, where a node which is not a part of any rescue group may pose as a registered member and hence claim authenticity, with several malicious intentions.
- ***Eavesdropping Attack***: In this attack, malicious nodes, while forwarding situational messages, discover the communication contents by listening to the data being transmitted, with a number of spiteful intentions.

– ***Modification Attack***: This is an attack where malicious nodes distort the transmitted messages either by wiping out some part of it or stuffing garbage into it.

5.2.1.3 Pin Distribution at the Set-up Phase

In post-disaster relief actions, volunteers work in groups, each of which has definite duties to carry out. Some of these groups, as identified by the International Federation of Red Cross and Red Crescent Societies [28] and Japan International Cooperation Agency [29], are: Search, Rescue and Evacuation Group, Basic Health Care Group, Logistics Group, Relief Group, Water and Sanitation Group, IT & Telecommunications Group, Disaster Response and Reconstruction Group. Thus, it is assumed that each volunteer should belong to such a group.

The control station assigns a group-id, say $GID_j, j = 1, 2, \ldots, n$, to each of the n groups. Additionally, the control station generates a pair of pins called group-pin (GP_j) and modified group-pin (MGP_j) for every group. The group-pin is in fact a private key created not for any individual volunteer but for the whole group. Likewise, the modified group-pin is in fact a public key shared by every member of a specific group. These group keys lessen the load of storing private key–public key pairs for every member; this is vital in view of the scarce resources (storage space, energy, etc.) of the devices that run the network. The group-pin for a group is kept secret to only the members of that group, and the modified group-pin is distributed among all registered members of all other groups and to all shelters as well. Finally, the control station creates a control node-id (*CID*) and a control-pin, modified control-pin pair (*CPin*, *MCPin*, respectively) for the control-node and distributes *CID* and *MCPin* to all forwarder-nodes. The list of pins stored with each type of node is presented in Table 5.1a–c.

When a new volunteer joins relief work, she registers with any one of these groups. On registration, the volunteer is provided a smartphone containing necessary ids and pins like *GID* of all groups, *GP* of its own group, *MGP* of other groups and *CPin*. Thus, new volunteers could be incorporated into the network without perturbing the existing network set-up. In the suggested schemes, the control station, which is a trusted authority, distributes keys to the groups at the set-up phase. The control station has absolutely no role in the run-time phase when nodes check authenticity of peers and provide integrity and confidentiality to messages in a entirely decentralized manner.

5.2.2 *Group-Based Distributed Authentication*

In this section, a group-based distributed authentication scheme is elaborated which authenticates nodes participating in data exchange, as members of some valid rescue groups by exchanging modified group-pins. Challenge–response authentication mechanism that uses a one-way hash function is used for these exchanges. Suppose

Table 5.1 (a) List of pins maintained at each shelter-node. (b) List of pins maintained at control-node. (c) List of pins maintained at each forwarder-node

Group-id	Group-pin	Modified group-pin
(a)		
GID_1	…	MGP_1
GID_2	…	MGP_2
…	…	…
GID_n	…	MGP_n
(b)		
GID_1	GP_1	MGP_1
GID_2	GP_2	MGP_2
…	…	…
GID_n	GP_n	MGP_n
CID	$CPin$	$MCPin$
(c)		
GID_1	–	MGP_1
GID_2	–	MGP_2
…	…	…
GID_{own}		GP_{own}
…	…	MGP_{own}
GID_n	–	MGP_n
CID	–	$MCPin$

a node, say FN_i, of group i wishes to forward a message to another node, say FN_j belonging to group j. Both FN_i and FN_j authenticate each other as registered members of valid rescue groups before beginning any message-forwarding activity. In this context, FN_i generates a random challenge string, say $Challenge(FN_i)$, and forwards it to FN_j. FN_j creates one more challenge string, say $Challenge(FN_j)$, and calculates the hash value of $MGP_j + Challenge(FN_i) + Challenge(FN_j)$ as:

$$H_j^j = H(\mathrm{MGP}_j + Challenge(FN_i) + Challenge(FN_j))$$

FN_j sends GID_j, $Challenge(FN_j)$ and H_j^j to FN_i. FN_i looks for the modified group-pin MGP_j related to GID_j in the maintained list of pins and computes the hash of $MGP_j + Challenge(FN_i) + Challenge(FN_j)$ as:

$$H_i^j = H(MGP_j + Challenge(FN_i) + Challenge(FN_j))$$

FN_i compares H_i^j and H_j^j. If the comparison is a success, FN_i validates FN_j as a registered member of group j and regards FN_j as an authentic forwarder. Next, FN_i computes a hash of $MGP_i + Challenge(FN_i) + Challenge(FN_j)$ as:

$$H_i^i = H(MGP_i + Challenge(FN_i) + Challenge(FN_j))$$

and sends GID_i and H_i^i to FN_j. FN_j looks for the modified group-pin MGP_i related to GID_i among the stored pins and calculates the hash value of MGP_i+$Challenge(FN_i)$+ $Challenge(FN_j)$ as:

$$H_j^i = H(MGP_i + Challenge(FN_i) + Challenge(FN_j))$$

FN_j compares H_j^i and H_i^i. If the comparison is a success, FN_j validates FN_i as an enrolled affiliate of group i and regards FN_i as a valid node.

In this group-based distributed authentication scheme, the modified group-pins are not forwarded as plain text and thus an attacker needs to brute force the hash function to find out the modified group-pin: an intimidating task. Thus, the scheme offers security to replay attacks. The above process authenticates two participating nodes as registered members of a rescue group and debars unauthorized nodes from participating in forwarding activities. Validating a node as a registered member of a valid rescue group lessens the burden of validating every node as an individual based on its personal credentials. This results in lesser time and resource (battery life, etc.) consumptions. The above scheme is explained in Algorithm 1.

Algorithm 1: Group Based Distributed Authentication Algorithm
Input : GID_j, $Challenge(FN_j)$, H_j^j
Output : GID_i, H_i^i, Decision about authenticity of FN_j
Runs At : Forwarder-node FN_i (DTN enabled smartphone, carried by volunteer)

Generate $Challenge(FN_i)$ and send it to FN_j
Receive GID_j, $Challenge(FN_j)$ and H_j^j from FN_j
Compute $H_i^j = H(MGP_j + Challenge(FN_i) + Challenge(FN_j))$
if $H_i^j = H_j^j$ **then**
return 1 // FN_j is a registered member of group j
else
return 0 // FN_j is not a registered member of group j
end if
Compute $H_i^i = H(MGP_i + Challenge(FN_i) + Challenge(FN_j))$
Send GID_i and H_i^i to FN_j

5.2.3 *Multilayer Hashed Encryption*

In this section, a multilayer hashed encryption scheme, inspired by the onion routing technique presented in [10], is explained. The concept is to encrypt a message with q ($\geq$2) layers of encryption with q modified group-pins so that an in-between forwarder-node can take away just one encryption layer with its own group-pin keeping the message protected by $q-1$ layers of encryption. The intermediate node verifies integrity of the encrypted message and decides whether to further forward the message. If the node finds that integrity has been maintained, it adds a new encryption layer with a different modified group-pin and forwards the message to the next node. In this fashion, no intermediate node can access the message in plain text. Also, all nodes use a common commutative encryption. The scheme is demonstrated below, with (q =) two layers of encryption.

5.2.3.1 Message Encryption at Shelter-Node

Suppose a shelter-node say SN_k, $k = 1, 2, \ldots, K$ creates a message $M_{k,p}$ (pth message from the kth shelter), containing resource demands or utilities of the corresponding shelter, and forwards it to one or more forwarder-nodes that move around the shelter. When SN_k intends to forward $M_{k,p}$ to FN_i, a forwarder-node working in its region, SN_k, first validates FN_i as a registered member of group i, using the group-based distributed authentication scheme, discussed above. Following this, SN_k encrypts $M_{k,p}$ with MGP_i as $E_{MGP_i}(M_{k,p})$. SN_k puts a second layer of encryption on $E_{MGP_i}(M_{k,p})$ with the modified group-pin MGP_j of any other group, say j as $E_{MGP_j}(E_{MGP_i}(M_{k,p}))$. SN_k then creates a hash of $E_{MGP_j}(E_{MGP_i}(M_{k,p}))$ as:

$$H_{i,j} = H(E_{MGP_j}(E_{MGP_i}(M_{k,p})))$$

and sends $[H_{i,j}, E_{MGP_j}(E_{MGP_i}(M_{k,p})), GID_j]$ to FN_i.

5.2.3.2 Message Verification at Forwarder-Node

On receiving each double encrypted message $E_{MGP_j}(E_{MGP_i}(M_{k,p}))$, FN_i computes a hash

$$H'_{i,j} = H(E_{MGP_j}(E_{MGP_i}(M_{k,p})))$$

and compares $H'_{i,j}$ with $H_{i,j}$. If the hashes fail to match, FN_i comprehends that the integrity of $M_{k,p}$ is missing and thus drops it. Or else, FN_i stores it for further forwarding. Like this, FN_i decides on whether to forward $M_{k,p}$ further or remove it from the network, without getting access to the message in plain text. FN_i finally decrypts $E_{MGP_j}(E_{MGP_i}(M_{k,p}))$ as $D_{GP_i}(E_{MGP_j}(E_{MGP_i}(M_{k,p})))$ to obtain $E_{MGP_j}(M_{k,p})$.

FN_i understands that there exists an encryption for group j through GID_j. FN_i stores the message either in a demand vector if it is a demand message or in a utility vector if it is a utility message. This way the demand and utility vectors of FN_i consist of encrypted demands and utilities of the shelter in which it works.

5.2.3.3 Message Sending at Forwarder-Node

When FN_i encounters FN_j, local or remote, they validate each other as a registered members of some rescue groups using the group-based distributed authentication scheme. If FN_i finds FN_j, of group j, as an authentic forwarder, it forwards the demand and utility vectors if FN_j is a local node and forwards the demand and utility matrices if FN_j is a node belonging to some remote region. FN_i adds another layer of encryption to each $E_{MGP_j}(M_{k,p})$ with the modified group-pin MGP_l of some other group, say l as $E_{MGP_l}(E_{MGP_j}(M_{k,p}))$. FN_i then calculates a hash

$$H_{j,l} = H(E_{MGP_l}(E_{MGP_j}(M_{k,p})))$$

and sends $[H_{j,l}, E_{MGP_l}(E_{MGP_j}(M_{k,p})), GID_l]$ to FN_j.

If FN_i finds FN_l belonging to group l as the next authentic forwarder-node, it adds another layer of encryption to $E_{MGP_j}(M_{k,p})$ with the modified group-pin MGP_l of the same group l, as $E_{MGP_l}(E_{MGP_j}(M_{k,p}))$. FN_i then creates a hash

$$H_{j,l} = H(E_{MGP_l}(E_{MGP_j}(M_{k,p})))$$

and sends $[H_{j,l}, E_{MGP_l}(E_{MGP_j}(M_{k,p})), GID_j]$ to FN_l.

Thus, the scheme enables the forwarder-nodes to perform secure comparison on an encrypted message to identify breach of integrity of the message and discard it accordingly. This early detection of damaged messages and their exclusion from the network saves considerable amount network resources. The above scheme is explained in Algorithm 2.

Algorithm 2: Multilayer Hashed Encryption Algorithm

procedure MESSAGE ENCRYPTION AT SHELTER-NODE

Input : $M_{k,p}$

Output : $H_{i,j}, E_{MGP_j}\left(E_{MGP_i}(M_{k,p})\right), GID_j$

Runs At : Shelter-node SN_k

: Encrypt $M_{k,p}$ as $E_{MGP_i}(M_{k,p})$

: Encrypt $E_{MGP_i}(M_{k,p})$ as $E_{MGP_j}\left(E_{MGP_i}(M_{k,p})\right)$

: Create $H_{i,j} = H(E_{MGP_j}\left(E_{MGP_i}(M_{k,p})\right))$

: Send $[H_{i,j}, E_{MGP_j}\left(E_{MGP_i}(M_{k,p})\right), GID_j]$ to FN_i

procedure MESSAGE VERIFICATION AT FORWARDER-NODE

Input : $H_{i,j}, E_{MGP_j}\left(E_{MGP_i}(M_{k,p})\right), GID_j$

Output : $E_{MGP_j}(M_{k,p})$, GID_j, Decision about integrity of $M_{k,p}$

Runs At : Forwarder-node FN_i (DTN enabled smartphone, carried by volunteer)

: Computes $H'_{i,j} = H(E_{MGP_j}\left(E_{MGP_i}(M_{k,p})\right))$

: **if** $H'_{i,j} == H_{i,j}$ **then** // Integrity of $M_{k,p}$ is preserved

: Obtain $E_{MGP_j}(M_{k,p})$ as $D_{GP_i}(E_{MGP_j}\left(E_{MGP_i}(M_{k,p})\right))$

: Store $E_{MGP_j}(M_{k,p})$ and GID_j for further forwarding

: **else** // Integrity of $M_{k,p}$ is lost

: Drop $M_{k,p}$

: **end if**

procedure MESSAGE SENDING AT FORWARDER-NODE

Input : $E_{MGP_j}(M_{k,p})$, GID_j

Output : $H_{j,l}, E_{MGP_l}\left(E_{MGP_j}(M_{k,p})\right), GID_l$ (or GID_j)

Runs At : Forwarder-node FN_i (DTN enabled smartphone, carried by volunteer)

: **if** FN_j is the next authentic forwarder **then**

: Encrypt $E_{MGP_j}(M_{k,p})$ as $E_{MGP_l}\left(E_{MGP_j}(M_{k,p})\right)$

: Compute $H_{j,l} = H(E_{MGP_l}\left(E_{MGP_j}(M_{k,p})\right))$

: Send $[H_{j,l}, E_{MGP_l}\left(E_{MGP_j}(M_{k,p})\right), GID_l]$ to FN_j

: **end if**

: **if** FN_l is the next authentic forwarder **then**

: Encrypt $E_{MGP_j}(M_{k,p})$ as $E_{MGP_l}\left(E_{MGP_j}(M_{k,p})\right)$

: Compute $H_{j,l} = H(E_{MGP_l}\left(E_{MGP_j}(M_{k,p})\right))$

: Send $[H_{j,l}, E_{MGP_l}\left(E_{MGP_j}(M_{k,p})\right), GID_j]$ to FN_l

: **end if**

5.2.4 *Performance Analysis*

The efficiency of the proposed multilayer encryption and hashing-based scheme discussed in the earlier section is enumerated through qualitative and quantitative analyses in the following subsections.

5.2.4.1 Qualitative Analysis

This section presents the performance of the scheme in terms of storage, computation and energy overheads. As mentioned in Sect. 5.2.1.1, shelter-nodes and the control-node are high-end computing devices like laptop, workstation and server. On the other hand, the forwarder-nodes that run the multilayer encryption and hashing-based scheme are DTN-enabled smartphones, which are limited in terms of processing power, energy and storage space. Thus, the storage, computation and energy necessities of this algorithm are enumerated to measure its suitability for resource-limited smartphones. The above overheads are computed for an operational period, which is normally 12 h [30] for the current application. In particular, the overheads are computed for the worst-case scenario that occurs if a particular forwarder-node encounters all other nodes in the network. These are the overheads accrued for the multilayer encryption and hashing-based scheme in addition to the overheads used for the opportunistic knowledge sharing algorithm. The Samsung Galaxy Grand 2 smartphone is considered for this analysis that has a Qualcomm Snapdragon processor with ARM core. Battery capacity is 2600 mAh with a voltage of 3.8 V and initial energy of 35,568 J [31].

In a DTN environment, as considered here, extensive brute-force attacks are less likely to be launched, due to the ephemeral and intermittent character of the network. In addition, the smartphones used for launching attacks are resource constrained and cannot launch computation-intensive cryptanalysis. According to the findings presented in [32] and [33], SHA-256 and 512-bit RSA are competent enough in terms of computation and energy costs. SHA-256 and 512-bit RSA offer adequate safety to a transient network, as ours, that remains alive for a short period until the normal communication starts running. Thus, SHA-256 is employed for hashing operations and 512-bit RSA is employed for encryption and decryption.

Storage Overhead

Storage overhead is measured in terms of the memory requirement for storing the different types of pins in the smartphones. In the proposed scheme, every forwarder-node stores n-modified group-pins related to the n groups, one group-pin of its own group and one control-pin of the control-node. Thus, every forwarder-node stores $(n + 2)$ pins. As 512-bit RSA is being employed, every pin is 512 bits long. As a

Table 5.2 Computation and energy overheads of multilayer encryption and hashing-based scheme

Process	Operation	Algorithm used	Computation overhead [ms]		Energy overhead [mJ]	
Sending message	Encryption	RSA—512	0.533	8.843	0.452	7.132
	Hashing	SHA—256	8.31		6.86	
Receiving message	Hashing	SHA—256	8.31	8.46	6.86	6.987
	Decryption	RSA—512	0.15		0.127	
Total			17.303		14.299	

result, storage requirement for maintaining $(n + 2)$ pins at each forwarder-node is $512 * (n + 2)$ bits.

Computation Overhead

Each forwarder-node carries out one encryption and one hashing for forwarding a message and carries out one hashing and one decryption after receiving a message. In the proposed scheme, since forwarder-nodes are not concerned with key generation, computational costs related to encryption and decryption only are counted. The computational costs, computed on the basis of results obtained in [32], are presented in Table 5.2.

Energy Overhead

The energy costs for the encryption, decryption and hashing occurring at each forwarder-node are computed on the basis of figures obtained in [32]. The results are presented in Table 5.2.

5.2.4.2 Quantitative Analysis

This subsection provides the quantitative analysis of the multilayer encryption and hashing-based scheme. The efficiency of the proposed scheme is evaluated through simulation.

Simulation Environment

An actual disaster situation is used for setting up the simulation environment, based on the 2015 Nepal earthquake. The Google Maps of water, food, shelter and medical resources for Nepal earthquake [34], depicted in Fig. 2.4, depicts the shelters and

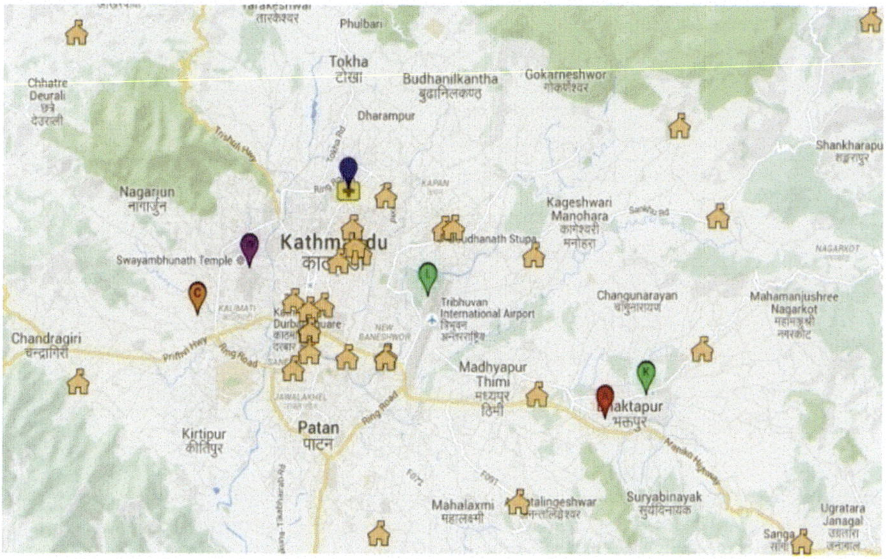

Fig. 5.1 Google Maps for the 2015 Nepal earthquake [34]

medical relief centres set in Kathmandu and its adjacent districts like Nuwakot, Sindhupalchowk and others (Fig. 5.1).

Simulation Set-up

The ONE simulator [35] is used to implement the scheme in a DTN environment. The simulation set-up used for this multilayer encryption and hashing-based scheme is almost similar to the set-up, explained in Sect. 3.4.2.2 of Chap. 3, used for the opportunistic knowledge sharing scheme. The map in [34] is used on ONE to plot nine shelters. It is assumed that ten forwarder-nodes, i.e. volunteers with smartphones, working in each shelter moving across the entire affected area, exchange situational messages of the shelters and reach them to the control station. The control station is set up at a distant location in the map. Information provided by an NGO, "Doctors For You" [36], who offered medical services during and after the earthquake, is used along with information available from [34, 37]. The shelter-nodes and the control-node are immobile. However, the forwarder-nodes are mobile smartphones and follow the post-disaster mobility (PDM) Model proposed by Uddin et al. in [38]. Post-disaster relief operations are usually carried out during daytime, typically from 6 a.m. to 6 p.m. Accordingly, the simulation time is kept as 12 h (43,000 s). A shelter-node creates situational messages at a rate of one per hour and broadcasts them to the forwarder-nodes. Since shelter requirements change infrequently, it is assumed that shelters generate one message per hour. The routing protocol used is spray and wait [39]. In the simulation, unauthorized, malicious and well-behaved nodes are

Table 5.3 Parameters used for simulation of multilayer encryption and hashing-based scheme

Parameter	Value
Number of shelters	9
Number of forwarder-nodes	90
Simulation period	12 h
Node speed	10–50 km/h
Transmission range	10 m
Message size	0.5 MB
Buffer size	5 MB
Message TTL	720 min
Movement model	PDM
Routing protocol	Spray and wait

deployed, who move in and around the disaster-affected area. Unauthorized nodes are those who do not have a valid group-pin and modified group-pin pair. Malicious nodes are those distort messages, while well-behaving nodes are normal nodes that take part in honest message forwarding. Table 5.3 specifies the important simulation parameters.

Simulation Metrics

The proposed multilayer encryption and hashing-based scheme is evaluated on the basis of the following metrics.

- ***Delivery Ratio***: It is the fraction of the messages delivered to destination nodes to those created by source nodes [40] defined as

$$\text{Delivery Ratio} = \frac{\text{No. of msgs. delivered}}{\text{No. of msgs. created}}$$

- ***Average Delay***: It indicates the average time a message takes from source to destination time [40] defined as

$$\text{Average Delay} = \frac{\sum_{\text{for all } M_D} (T_D - T_S)}{\text{No. of msgs. delivered}}$$

  where T_D is the time when message reaches its destination and T_S is the time when the message was created at source.
- ***Overhead Ratio***: It indicates the ratio of the number of control messages (including route request/reply/update/error packets) to the number of data messages [40] defined as

$$\text{Overhead Ratio} = \frac{\text{No. of control packets relayed}}{\text{No. of data packets relayed}}$$

- ***Attraction Ratio***: It indicates the rate at which messages are forwarded to selfish nodes [2] defined as

$$\text{Attraction Ratio} = \frac{\text{No. of msgs. received by selfish nodes}}{\text{No. of msgs. relayed}}$$

Results and Discussion

The proposed multilayer encryption and hashing-based scheme (running on top of the opportunistic knowledge sharing scheme) is compared with a simple spray and wait protocol in terms of the performance metrics defined above. The results shown here are the mean of 50 simulation runs.

- ***Impact on Delivery Ratio***: Figure 5.2a depicts the change in delivery ratio with rising number of unauthorized and malicious nodes. It seems that the proposed scheme achieves 60.67% higher delivery ratio.
- ***Impact on Average Delay***: Figure 5.2b shows the change in average delay with growing number of unauthorized and malicious nodes. It seems that the proposed scheme suffers only 8.77% increase in latency.
- ***Impact on Overhead Ratio***: Figure 5.2c shows the change in overhead ratio with mounting number of unauthorized and malicious nodes in the network. It seems that the proposed scheme achieves 62.98% lower overhead ratio. With rising number of malevolent nodes, the performance steadily degrades, but overhead ratio with the proposed scheme is still lower.
- ***Impact on Attraction Ratio***: Figure 5.2d shows the attraction ratio with a growing number of unauthorized and malicious nodes. It seems that the proposed scheme achieves 45.1% lower attraction ratio.
- ***Impact of Number of Encryption Layers on Average Delay***: Security of the proposed scheme relies on q, the number of encryption layers. If $q = 2$, two neighbouring nodes may conspire and decrypt the message completely. A bigger value for q requires a larger number of nodes to conspire to break the security of the message. Figure 5.3 shows that there is only 1.25% increase in average delay if number of encryption layers is raised from 2 to 5.

In this work, a group-pin-based mechanism for mutual authentication of nodes and a multilayer encryption and hashing-based scheme for providing full privacy and integrity to sensitive situational messages are proposed. In this scheme, in the event of a node capture, an attacker will not be able to access the message as the message at anytime is protected by $q - 1$ layers of encryption. Here, groups collectively add towards maintaining the confidentiality and integrity of forwarded messages (i.e. emergency resource needs). This scheme assists a node in identifying the integrity of a message at every hop without getting access to its substance in plain text and

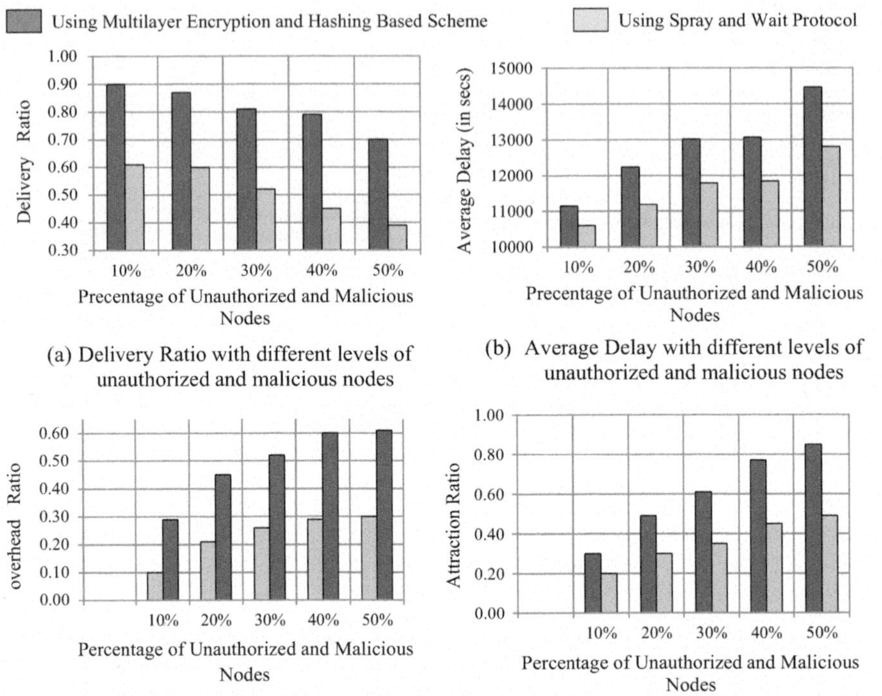

(a) Delivery Ratio with different levels of unauthorized and malicious nodes

(b) Average Delay with different levels of unauthorized and malicious nodes

(c) Overhead Ratio with different levels of unauthorized and malicious nodes

(d) Attraction Ratio with different levels of unauthorized and malicious nodes

Fig. 5.2 Performance of the multilayer encryption and hashing-based scheme with respect to different performance metrics

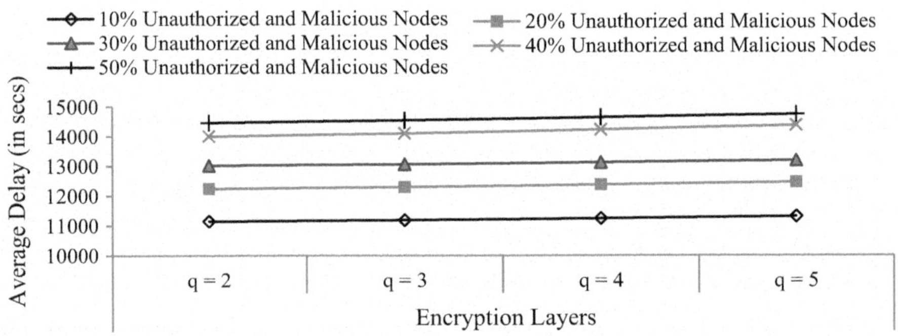

Fig. 5.3 Impact of increasing number of encryption layers on average delay

hence in taking decision on whether to keep on forwarding the message further. This facilitates timely identification of corrupt messages and their exclusion from the network without wasting additional network resources. Simulation results show that the scheme tenders 60.67% higher delivery ratio, 62.98% lower overhead ratio and 45.1% lower attraction ratio with just 8.77% increase in average delay.

After this, a security-aided and group encounter-based PRoPHET routing protocol is presented for secured transmission of categorized resource needs from the shelters to the relief camps. Message transmission is achieved through encounters among volunteers of different relief groups. Security features are integrated into the DTN routing protocol PRoPHET to offer security against privacy attack like eavesdropping and communication attacks like black hole, identity spoofing and bundle store overflow.

5.3 SAGE-PRoPHET: Security-Aided and Group Encounter-Based PRoPHET Routing Protocol

In this section, a security-aided and group encounter-based PRoPHET routing protocol [41, 42] is proposed for secured dissemination of categorized resource needs from various shelters to the relief camps. Categorization is done by the forwarder-nodes, on-the-fly, without compromising on their confidentiality to expedite the delivery of such categorized needs to their appropriate destinations. Message forwarding is accomplished on the basis of encounters among volunteers of different relief groups. Security features are incorporated into the DTN routing protocol PRoPHET to provide security against privacy attack like eavesdropping and communication attacks like black hole, identity spoofing and bundle store overflow.

As mentioned in Sect. 5.2.1.3, disaster management activities in a post-disaster scenario are conducted by various rescue groups, each group having explicit and more or less distinct tasks to execute. Thus, every volunteer belongs to one such group. For example, volunteers belonging to the medical group operate from a medical camp to cater to medical needs of the victims, whereas volunteers of the search evacuation group handle rescue and relocation activities from a rescue camp. Volunteers belonging to different rescue groups relay situational messages containing the resource needs of the shelters. These resources are of different categories, such as medical resources and food resources, and hence their needs are to be transmitted to the corresponding relief camps for necessary action. Thus, it becomes crucial for a forwarder-node (volunteer with smartphone) to categorize a received message, on-the-fly, according to their content, to identify the relief camp to which the message is to be delivered to. However, due to possible eavesdropping by malicious nodes in the network, content of sensitive situational messages cannot be made accessible even if that helps in categorization. Therefore, a secured message categorization technique that enables forwarder-nodes to categorize situational messages without compromising on their confidentiality needs to be proposed.

In a usual emergency response scenario, volunteers of a particular group once in a while go to their respective camps for replenishing relief material, etc. Thus, these volunteers are the most appropriate carriers of messages intended for that specific camp. However, owing to group dynamics, a volunteer of a specific group meets volunteers belonging to its own group or some specific groups more often than volunteers of any other groups. Hence, it is logical to utilize these meeting opportunities for transmitting categorized situational messages to their intended camps. PRoPHET [43], one of the yardstick DTN routing protocols, suits such encounter history-based routing because it makes use of the history of earlier encounters. However, to use PRoPHET for a group encounter-based routing, it needs tuning with the group movement patterns and history of group encounters.

PRoPHET depends upon inherent supposition that every node is truthful and assists in message forwarding. This supposition turns out wrong in the presence of malicious nodes that can instigate severe privacy attack like eavesdropping and communication attacks like black hole, identity spoofing and bundle store overflow. Thus, integrating security elements into PRoPHET for securing the network as well as the transmitted messages is imperative.

In this work, a security-aided and group encounter-based PRoPHET routing protocol, called SAGE-PRoPHET, is proposed. First, a public key encryption with keyword search (PEKS) [20]-based message categorization technique is suggested which is used by the forwarder-nodes to categorize messages at the run-time according to their content, without getting access to the message content. Next, PRoPHET is adapted for group encounter-based message forwarding to forward categorized messages, for a particular relief camp, through volunteers of that group or who encounter members of that group frequently to speed up the delivery of such messages. Finally, security features, which use group-based keys, are incorporated into PRoPHET for protecting the messages from eavesdropping attack and network from identity spoofing and bundle store overflow attacks. The proposed protocol handles eavesdropping attack through message encryption, bundle store overflow attack through message signing, identity spoofing attack through key encryption and black hole attack through encounter tokens.

5.3.1 System Model

In this section, the system model related to proposed protocol is described. The system model consists of the hierarchical structure of disaster relief operations, network architecture, routing protocol, attack model and pin and keyword distribution at the set-up phase.

5.3.1.1 Typical Hierarchy of Disaster Management Operations

In a post-disaster situation, rescue groups involved in relief activities set up relief camps in the disaster-affected area [28, 29]. Depending on their roles, relief organizations establish various kinds of camps like healthcare camps, relief camps and disaster response camps. Every such camp consists of some devoted volunteers who offer particular kind of service relevant to their camp. For example, volunteers registered with the Health Care Camp offers healthcare and those registered with the Relief Camp offers food, clothes, blankets, etc. A control station is established, at a distant location, for conducting disaster management operations in the camps [44, 45]. Therefore, post- disaster management tasks are executed hierarchically, led by the control station that monitors the functioning of different relief camps, which organize the actions of the volunteers associated with that specific camp. Such hierarchical system fosters collective and synchronized endeavour for disaster management. The typical hierarchy of disaster management operations is depicted in Fig. 5.4.

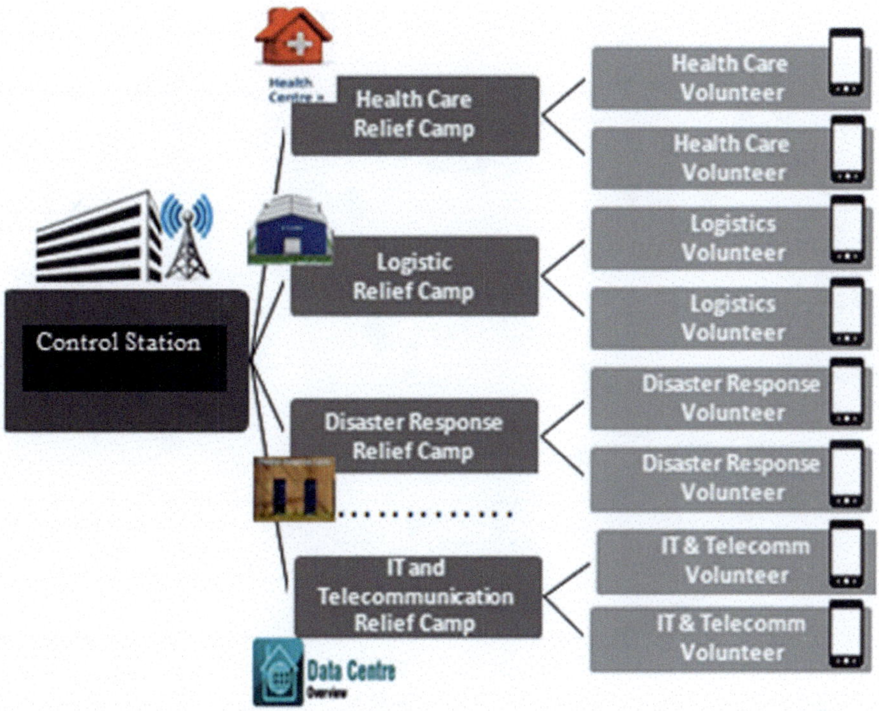

Fig. 5.4 Typical hierarchy of disaster management operations [41]

5.3.1.2 Network Architecture

The underlying network architecture of the proposed SAGE-PRoPHET protocol is mostly the same as the network architecture used for the multilayer encryption and hashing-based scheme, illustrated in Sect. 5.2.1.1. In this work, apart from shelter-nodes, control-node and forwarder-nodes, another type of node, called the camp-node, is introduced. A *camp-node* (e.g. a laptop or a workstation), housed in a group relief camp, receives categorized messages containing shelter resource needs (relevant to that camp) from the volunteers registered with that camp. The camp-nodes are equipped with Internet facilities, using which the shelter resource needs are transmitted to the control station. The camp-nodes act as intermediate drop boxes for collecting categorized resource needs. This eliminates the necessity of volunteers travelling to the control station for delivering shelter resource needs. A typical emergency communication network, composed of shelter-nodes, forwarder-nodes, camp-nodes and control-node, is illustrated in Fig. 5.5.

Fig. 5.5 An emergency communication network with shelter-nodes, forwarder-nodes, camp-nodes and control-node [41]

5.3.1.3 Routing Protocol

The forwarder-nodes exchange messages among themselves following the DTN routing protocol PRoPHET [43]. This protocol uses the history of previous encounters and the transitive property to estimate a delivery predictability (*DP*), $P(A, B)$, at each node A for all known destinations B, kept in a *DP* table. As nodes meet, they exchange *DP* tables and update their own *DP* table according to the following equations. For an encountered node B, node A updates its *DP* using the following equation

$$P(A, B)_{new} = P(A, B)_{old} + (1 - P(A, B)_{old}) \times P_{enc}$$

where P_{enc} is a configurable parameter. Due to the transitive property of *DP*, A updates *DP*s for all other destinations i known by B based on the *DP* table provided by B. Transitive update is done using the following equation where β is a configurable parameter

$$P(A, i)_{new} = \max(P(A, i)_{old}, P(B, i) \times P(A, B)_{new} \times \beta)$$

In order to eliminate stale information from the network, the *DP* table is periodically aged according to the following equation for all destinations i, where γ is a constant and T is the number of time units since the last ageing process.

$$P(A, i)_{new} = P(A, i)_{old} \times \gamma^T$$

In PRoPHET, when a node A meets another node B, a message is forwarded to B only if the delivery predictability of the destination of the message is higher at B.

5.3.1.4 Attack Model

Serious security attacks can be instigated on PRoPHET. Out of these, the threat model considered in this work consists of the following four attacks:

- ***Eavesdropping Attack***: A malicious node, while forwarding situational messages, discovers the communication contents by listening to the data being transmitted, with several malicious intentions.
- ***Identity Spoofing Attack***: A spiteful entity feigns a different identity. This is used to embezzle messages that are destined to a specific node.
- ***Black hole Attack***: A spiteful node, behaving as a black hole, assigns a high (nearly 1) delivery predictability for every node in the network, attracts all messages from the nodes it encounters and forwards them to none. An ignorant node meeting a black hole forwards all messages to it, with a belief that the messages will be truthfully forwarded. Owing to the transitivity property, delivery predictabilities presented by these black holes will influence the delivery predictabilities of all other forwarders.

– ***Bundle Store Overflow Attack***: A spiteful forwarder might create a huge number of false messages for a specific node and clog the buffer of that targeted node, at the cost of genuine messages. Furthermore, these spiteful nodes might redistribute stale messages, which make the situation worse by storage run over if the messages are relayed before their expiry.

5.3.1.5 Pin and Keyword Distribution at the Set-up Phase

In the proposed protocol, N forwarder-nodes (FN), S shelter-nodes (SN) along with n camp-nodes (CN), pertaining to n groups, are assumed. The control station allots a group-id, say GID_j, $j = 1, 2, \ldots, n$, to each of the n groups. On top of this, the control station generates a group-pin (GP_j), modified group-pin (MGP_j) pair for each group. The group-pin is practically a private key created for the entire group. Similarly, the modified group-pin is in fact a public key common for all members of a group. These group keys lessen the load of storing keys for every volunteer, a very important consideration owing to the restricted resources of the smartphones that run the network. GP_j remains undisclosed inside the members of that group, whereas GID_j and MGP_j are shared with all. Every shelter-node is assigned a shelter-id, say SID_s, $s = 1, 2, \ldots, S$ and a shelter-pin, modified shelter-pin pair, ($SPin_s$, $MSPin_s$, respectively). $MSPin_s$ is shared with all camp-nodes and registered forwarder-nodes. Finally, every camp-node is allotted a camp-id (CID_j) and a camp-pin, modified camp-pin pair, ($CPin_j$, $MCPin_j$, respectively). CID_j and $MCPin_j$ are distributed to all shelter-nodes. A new entrant can join relief operations by registering with a group. After registration, the volunteer is given a smartphone that consists of all necessary ids and pins. Therefore, new volunteers are easily incorporated into the network.

In addition to such pins, the control station provides lists of prescribed keywords to the shelter-nodes and forwarder-nodes. The keywords are so chosen that each of them relate to a specific service group. Examples of such keywords may be stranded, evacuation, etc., related to the Search, Rescue and Evacuation Groups; vaccination, first aid, ambulance, penicillin, etc., related to the Health Care Groups; rice, milk, bedding, clothing, light, plastics, etc., related to the Logistics Groups; bridge, building, roads, etc., related to the Reconstruction Groups, and likewise. Thus, the control station provides lists of keywords $\{K_m\}$, $m = 1, 2, \ldots, M$, classified by their types, as shown in Table 5.4, to shelter-nodes and forwarder-nodes. Also, each forwarder-node stores within it a priority queue PQ_j corresponding to each group $j, j = 1, \ldots, n$ for storing messages related to that group.

In the proposed protocol, the control station is a trusted authority that distributes keys and classified lists of keywords at the set-up phase. Afterwards, the control station has no role during the run-time when forwarder-nodes transmit messages using the secured PRoPHET routing protocol in a completely decentralized manner.

Table 5.4 List of keywords classified by service types stored at shelter-nodes and forwarder-nodes

Category	Keywords
Healthcare keywords	$K_1, K_2, \ldots, K_{p-1}, K_p$
Logistics keywords	$K_{p+1}, K_{p+2}, \ldots, K_{r-1}, K_r$
…	… … … … … … … … … …
Search and rescue keywords	… … … … … …, K_{M-1}, K_M

5.3.2 *Categorizing Situational Messages Using PEKS*

In this section, the public key encryption with keyword search (PEKS)-based message categorization technique is proposed. The technique classifies messages received by a forwarder-node depending on the keywords contained in the message header, without allowing the node to access the content of the message. The PEKS method as proposed by Boneh et al. [20] is explained first, followed by the proposed technique.

5.3.2.1 Public Key Encryption with Keyword Search

PEKS allows an intermediate node to search for a keyword in an encrypted message that a sender has sent to a receiver. The intermediate node learns nothing except whether the keyword occurs in the message or not. The technique is illustrated with three nodes, say A, B and C. If B wants to send a message M with keywords $W_1, W_2, \ldots, W_k$ to A through C, B encrypts M using a standard public key encryption technique and then appends to the resulting ciphertext, a *PEKS* of each keyword. Thus, B computes and sends

$$\left[E_{A_{pub}}(M), PEKS\left(A_{pub}, W_1\right), \ldots, PEKS\left(A_{pub}, W_k\right)\right]$$

to A through C, where A_{pub} is the public key of A. For a public key A_{pub} and a keyword W, a searchable encryption of W

$$S = PEKS\left(A_{pub}, W\right)$$

enables searching for specific keywords without revealing any information about the message. C, the intermediate node, cannot infer any information on M at this step. If A wants to enable C to search for the keyword W in the messages that are addressed to it, A has to provide C with a trapdoor

$$T_W = TRAPDOOR\left(A_{priv}, W\right)$$

where A_{priv} is the private key of A. Given A_{pub}, S and a trapdoor T_W, C proceeds to search for the keyword W'' in M using the test function

$$TEST(A_{pub}, S, T_W)$$

that outputs true if $W = W'$ and false otherwise. If $W = W'$, C only knows that M contains the keyword W, and if the result is false it just deduces that M does not contain W.

5.3.2.2 Message Categorization Using PEKS

The proposed technique uses *PEKS* to encrypt shelter messages and their headers and allow intermediate forwarder-nodes to search for keywords in the encrypted header to enable them categorizing the messages.

- ***Trapdoor Distribution Phase***: In the set-up phase, the control station provides each forwarder-node a trapdoor corresponding to each prescribed keyword K_s, $s = 1, 2, \ldots, m$, as

  $$T_{K_s} = TRAPDOOR(CPin, K_s)$$

  where $CPin_j$, $j = 1, 2, \ldots, n$, is the camp-pin of the jth camp-node. Therefore, each forwarder-node carries with it M such trapdoors.
- ***Header Generation and Encryption Phase***: Whenever a shelter-node, say SN_t, $t = 1, 2, \ldots, K$, generates a message $MSG_{t,n}$ (nth message from the tth shelter) providing shelter information, it creates a header for the message

  $$H\left(MSG_{t,n}\right) = \{K_s\}, s = 1, \ldots, m$$

  containing a set of prescribed keywords extracted from the message content, where $m \subset M$, the total number of keywords prescribed by the central control station. In order to protect privacy, SN_t encrypts $H\left(MSG_{t,n}\right)$ using a standard public key encryption technique and then appends to the resulting ciphertext, a PEKS of each keyword. Thus, SN_t computes

  $$H'(MSG_{t,n}) = E_{MCPin}(H(MSG_{t,n}))||PEKS(MCPin, K_1)||\ldots||PEKS(MCPin, K_m)$$

  where $MCPin$ is modified camp-pin of the camp-node.
- ***Message Encryption Phase***: In order to provide complete privacy, SN_k encrypts $M_{t,n}$ using a standard public key encryption technique and $MCPin$. Thus, SN_k computes

  $$MSG'_{t,n} = E_{MCPin}\left(MSG_{t,n}\right)$$

Table 5.5 Keyword frequency distribution for $MSG_{t,n}$ generated by *FN*

Group-id	Frequency of keywords
GID_1	$FREQ_1(MSG_{t,n})$
GID_2	$FREQ_2(MSG_{t,n})$
...	
GID_n	$FREQ_n(MSG_{t,n})$

Finally, SN_t concatenates the encrypted header with the encrypted message to form an augmented message

$$A(MSG_{t,n}) = H'\left(MSG_{t,n}\right)||MSG'_{t,n}$$

and sends it to an appropriate forwarder-node that comes in contact with it.

- ***Keyword Search Phase***: When a forwarder-node, say *FN*, receives the message $A(MSG_{t,n})$, either from a shelter or from another forwarder-node, it proceeds to search for each of the prescribed keywords K_s, in the encrypted message header $H'\left(MSG_{t,n}\right)$. Given *MCPin*, $PEKS(MCPin, K_s)$ and a trapdoor T_{K_s}, *FN* can search for the keyword K_s' in $H'\left(MSG_{t,n}\right)$ using the *TEST* function

$$TEST(MCPin, PEKS(MCPin, K_s))$$

that outputs true if $K_s = K'_s$ and false otherwise. If K_s, K'_s, *FN* knows that $MSG_{t,n}$ contains the keyword K_s, and deduces that $MSG_{t,n}$ does not contain K_s otherwise.
- ***Message Categorization Phase***: The forwarder-node *FN* executes the *TEST* function on the message header $H'\left(MSG_{t,n}\right)$ with each of the prescribed keywords, as stored in Table 5.4, and keeps a count of the number of keywords associated with each group in the message. If $FREQ_j(MSG_{t,n})$ denotes the frequency of keywords in $MSG_{t,n}$ associated with the group $j, j = 1, 2, \ldots., n$, then *FN* generates a frequency distribution as shown in Table 5.5.

FN finds the mode of the above frequency distribution, i.e. the group-id for which frequency is highest. If $FREQ_i(MSG_{t,n})$, $i = 1, \ldots, n$, is highest, then it becomes evident that the message contains more information for the group i.

Now, in an opportunistic network like ours, as forwarders have very little contact period with each other, only a few messages can be transferred per contact. Thus, messages that are of high priority and high relevance to a receiver must be forwarded before any other message. To accomplish this, *FN* inserts $MSG_{t,n}$ in the priority queue PQ_i designated for group i. While inserting, *FN* puts $MSG_{t,n}$ in its appropriate position so that the prioritized nature of PQ_i is maintained; i.e. if for any other message $MSG_{p,r}$ (rth message from the pth shelter) $FREQ_i(MSG_{p,r}) > FREQ_i(MSG_{t,n})$, then $MSG_{p,r}$ should be placed before $MSG_{t,n}$ so that $MSG_{p,r}$ gets delivered first. This way, high-priority messages are transmitted at the first possible opportunity.

If the frequency distribution below turns out to be bimodal, i.e. both $FREQ_i(MSG_{t,n})$ and $FREQ_j(MSG_{t,n})$, $(i, j) = 1, \ldots, n)$, are highest, then certainly the message is of relevance to both the groups i and j. Then, FN inserts $MSG_{t,n}$ in PQ_i as well as in PQ_j. Situations where there are more than two modes are handled similarly.

– ***Message Decryption Phase***: Finally, the camp-node on receiving a message $A(MSG_{t,n})$ decrypts the encrypted message $MSG'_{t,n}$ as

$$D_{CPin}(E_{MCPin}(MSG_{t,n}))$$

to obtain the original message $MSG_{t,n}$ and allocates relief stock for shelter SN_t according to the requirements specified in $MSG_{t,n}$.

The above technique enables a forwarder-node to categorize received messages and put them in appropriate priority queues without learning anything about the content of the messages.

5.3.3 Adapting PRoPHET for Group Encounter-Based Routing

In this section, the PRoPHET routing protocol is adapted for group encounter-based message forwarding from various shelters to the relief camps, by utilizing history of group encounters instead of personal encounters. For this, a group-DP table is created and maintained by every node, keeping the delivery predictability of every node, with all rescue groups operating in the network. $P\big(FN_i, GID_j\big)$ is defined as the group delivery predictability of a forwarder-node FN_i, of group i, with a member of group j. Every time FN_i, of group i, and FN_j of group j meet with each other, they interchange their group-DP tables and modify their tables on the basis of the group-ids and the received table. For an encountered node FN_j, node FN_i modifies its group delivery predictability using the following equation

$$P(FN_i, GID_j)_{new} = P(FN_i, GID_j)_{old} + (1 - P\big(FN_i, GID_j)_{old}\big) \times P_{enc}$$

Likewise, for an encountered node FN_i, node FN_j revises its group-DP by means of the following equation

$$P(FN_j, GID_i)_{new} = P(FN_j, GID_i)_{old} + (1 - P\big(FN_j, GID_i)_{old}\big) \times P_{enc}$$

Note that, $P\big(FN_i, GID_j\big)$ is different from $P\big(FN_j, GID_i\big)$ because FN_i might not have encountered that many number of volunteers from group j, as it has met volunteers of group i. FN_i modifies transitive group-DPs for all other groups k met by FN_j with the equation

$$P(FN_i, GID_k)_{new} = \max\left(P(FN_i, GID_k)_{old}, P\left(FN_j, GID_k\right) \times P\left(FN_i, GID_j\right)_{new} \times \beta\right)$$

FN_j revises transitive group-DPs for every other group k encountered by FN_i exploiting the equation

$$P\left(FN_j, GID_k\right)_{new} = \max\left(P\left(FN_j, GID_k\right)_{old}, P(FN_i, GID_k) \times P\left(FN_j, GID_i\right)_{new} \times \beta\right)$$

For eliminating out-of-date messages from the network, the group-DP table is passed through ageing at regular intervals for every group k through the equation

$$P(FN_i, GID_k)_{new} = P(FN_i, GID_k)_{old} \times \gamma^T$$

Using the above equations, PRoPHET can send categorized messages, intended for a specific camp, through volunteers of that group or who meet volunteers of that group frequently, for speeding up delivery of these messages. In group encounter-based PRoPHET, a node stores delivery predictabilities with groups rather than delivery predictabilities with individual node in the network. This lessens the burden of computing and maintaining delivery predictabilities for all nodes.

5.3.4 *Incorporating Security Features into PRoPHET*

Once PRoPHET is adapted to probabilistically send messages using history of group encounters, the next task is to incorporate security elements into the protocol for secured distribution of situational messages. In this section, the threats pointed out in Sect. 5.3.1.4 are addressed to immune PRoPHET against possible attacks by spiteful entities.

5.3.4.1 Message Encryption at Shelter-Node

Suppose shelter-node with shelter-id SID_k, $k = 1, \ldots, K$ generates a message M destined to camp of the jth group, $j = 1, \ldots, n$, with id CID_j at time point T_t. For offering end-to-end privacy to situational messages, SID_k encrypts M with $MCPin_j$ as

$$E_{MCPin_j}(M)$$

in order that just the camp-node of the jth group should be able to decrypt the message with $CPin_j$ as

$$D_{CPin_j}\left(E_{MCPin_j}(M)\right)$$

to obtain M. Therefore, no malicious node is able to snoop messages targeted to the camps and thus privacy of susceptible situational messages is conserved. Control messages and disaster management commands issued by the control station for different rescue groups are encrypted using modified group-pins in order that volunteers of the appropriate group should be able to decrypt them.

5.3.4.2 Message Signing at Shelter-Node

For controlling bundle store overflow attack, SID_k produces a signature S_k to digitally sign a message with its shelter-pin $SPin_k$. Thus,

$$S_k = E_{SPin_k}\left(H\left(E_{MCPin_j}(M), SID_k, CID_j, T_t\right)\right)$$

where H is a hash function. SID_k transmits the expanded message M_k^j, for CID_j. Thus,

$$M_k^j = \left\{E_{MCPin_j}(M), SID_k, CID_j, T_t, S_k\right\}$$

M_k^j gets forwarded to the relief camp CID_j either through volunteers of group j or volunteers who encounter volunteers of group j often.

5.3.4.3 Key Encryption at Forwarder-Node

To counteract the identity spoofing attack by illegal nodes, when two forwarder-nodes meet each other, they validate each other prior to exchanging messages. So, when FN_i of group i aims to transmit a message to FN_j of group j, they are required to validate one another as legitimate volunteers of authorized groups. As the MGPs are shared with just the registered volunteers (and not to non-registered intruders), deducing validation from ownership of pins is reasonable. Thus, both FN_i and FN_j request their *GID*s and *MGP*s to validate each other. However, *MGP*s cannot be interchanged openly as plain text as illegal nodes may pry these values and use them in the future to allege legitimacy. A key encryption technique is used to address this issue. Here, FN_j encrypts MGP_j with the GP_j as

$$E_{GP_j}(MGP_j)$$

This encrypted pin as well as GID_j [i.e. GID_j, E_{GP_j}, (MGP_j)] is then forwarded to FN_i. The GP_j, MGP_j pair are so created that messages encrypted with GP_j can be decrypted with MGP_j and vice versa. Thus, FN_i, who owns MGP_j of group j, decrypts

$$E_{GP_j}(MGP_j) \text{ as } D_{MGP_j}\left(E_{GP_j}(MGP_j)\right)$$

to get MGP_j. FN_i contrasts MGP_j with the stored MGPs for each and every group. If the decrypted MGP matches with the stored MGP, FN_i validates FN_j as a legitimate member of group j and therefore selects FN_j as a forwarder. Analogous process is used by FN_j to authenticate FN_i as a valid affiliate of group i. As the modified group-pin MGP_j is shared only with registered volunteers, attackers of the network are unable to decrypt the pin without the knowing MGP_j.

These mechanisms fruitfully validate two forwarder-nodes as affiliates of legitimate groups and prohibit unlawful nodes from participating in message transmission, without involving a central authority. Furthermore, authenticating a node as a legitimate member of a group lessens the overhead of authenticating every node individually.

5.3.4.4 Message Verification at Forwarder-Node

A forwarder-node, prior to storing the message M_k^j, decrypts S_k using $MSPin_k$ as

$$D_{MSPin_k}\left(E_{SPin_k}\left(H\left(E_{MCPin_j}(M), SID_k, CID_k, T_t\right)\right)\right)$$

to get a decrypted hash

$$H\left(E_{MCPin_j}(M), SID_k, CID_k, T_t\right)$$

The node calculates a hash of $\left\{E_{MCPin_j}(M), SID_k, CID_k, T_t\right\}$ as

$$H' = H\left(E_{MCPin_j}(M), SID_k, CID_k, T_t\right)$$

and contrasts H' against the decrypted hash. If the hashes are equal, FN_i believes that the message is from SID_k and stores it. This facilitates verification of the origin of the message. Hence, malicious forwarders are unable to introduce false messages in the network and clog the storage of a forwarder-node.

Additionally, malicious forwarders are unable to repercolate stale messages from valid shelters to a volunteer node and clog it, as every message has its creation timestamp. A forwarder-node, before storing a message in its buffer, verifies its timestamp and authenticity.

5.3.4.5 Encounter Tokens

- ***Encounter Token Generation by Forwarder-Node***: Digitally signed encounter tokens are used to thwart black hole attacks so as to curb spiteful nodes in drawing messages from other nodes by assigning elevated delivery predictabilities for every node. For this, whenever FN_i meets FN_j, a fresh encounter evidence

$$\{FN_i, FN_j, GID_i, GID_j, T_{ts}\}$$

is created as the evidence of the encounter of FN_i belonging to group i and FN_j belonging to group j at time point T_{ts}. FN_i creates a signature S_i to digitally sign the encounter evidence with GP_i. Thus,

$$S_i = E_{GP_i}(H(FN_i, FN_j, GID_i, GID_j, T_{ts}))$$

and creates a digitally signed encounter token $E_{i \to j}$ for FN_j, where

$$E_{i \to j} = \{FN_i, FN_j, GID_i, GID_j, T_{ts}, S_i\}$$

FN_i gives $E_{i \to j}$ to FN_j. In the same way, FN_j creates a signature S_j to digitally sign the encounter evidence with GP_j as

$$S_j = E_{GP_j}(H(FN_i, FN_j, GID_i, GID_j, T_{ts}))$$

and creates a digitally signed encounter token $E_{j \to i}$ for FN_i, where

$$E_{j \to i} = \{FN_i, FN_j, GID_i, GID_j, T_{ts}, S_j\}$$

FN_j gives $E_{j \to i}$ to FN_i. Hence, FN_j has a digitally signed encounter token from an affiliated volunteer of group i that is utilized as a proof of encountering a volunteer of that group. The number, $N(E_{i \to j})$, of digitally signed encounter tokens represents the number of encounters with members of group i. Likewise, FN_i collects such tokens from the volunteers of group j.

Similarly, FN_i owns a digitally signed encounter token from an affiliated volunteer of group j that can be used as evidence of encountering a volunteer from that group. The number, $N(E_{j \to i})$, of digitally signed encounter tokens shows the number of encounters with members of group j.

- ***Encounter Token Verification by Forwarder-Node***: When FN_k, of group k, wishes to forward a message for the ith group camp-node having camp-id CID_i via FN_j, it checks the group-DP $P(FN_j, GID_i)$ and other information to gauge FN_j as an apposite forwarder of that message. FN_k also checks the digitally signed encounter tokens $E_{i \to j}$, as evidences of encounters of FN_j with members of group i that FN_j has collected since their last meeting. FN_k decrypts S_i for each $E_{i \to j}$ with MGP_i as

$$D_{MGP_i}\left(E_{GP_i}\left(H\left(FN_{i,}FN_{j,}(M), GID_i, GID_j, T_{ts}\right)\right)\right)$$

to acquire a decrypted hash $H\left(FN_{i,}FN_{j,}(M), GID_i, GID_j, T_{ts}\right)$. FN_k calculates a hash of $\{FN_{i,}FN_{j,}(M), GID_i, GID_j, T_{ts}\}$ as

$$H' = H\left(FN_{i,} FN_{j,} (M), GID_i, GID_j, T_{ts}\right)$$

and compares H' with the decrypted hash. If the hashes turn out to be the same, FN_k believes that the encounter token is real. In similar fashion, FN_k authenticates all encounter tokens and counts the number of such tokens.

A large value of $P\left(FN_{j,} GID_i\right)$, close or equal to 1, asserts that FN_j has encountered volunteers of group i regularly, since its last encounter with FN_k. But, as a malicious node acting as a black hole might set its delivery predictabilities for all destinations to a very high value, such an assertion has to be backed by $N(E_{i \to j})$, the actual number of encounters with volunteers of group i. Thus, FN_k judges $FN_{j,}$ on the basis of the following rules:

Case *I*:

If $P\left(FN_{j,} GID_i\right) > P(FN_{k,} GID_i)_{old}$ and $N(E_{i \to j}) > N(E_{i \to k})$, then FN_j is an appropriate forwarder of messages directed to CID_i.

Case *II*:

If $P\left(FN_{j,} GID_i\right) < P(FN_{k,} GID_i)_{old}$ and $N(E_{i \to j}) < N(E_{i \to k})$, then FN_j is not an appropriate forwarder of messages directed to CID_i.

Case *III*:

If $P\left(FN_{j,} GID_i\right) > P(FN_{k,} GID_i)_{old}$ and $N(E_{i \to j}) < N(E_{i \to k})$, then FN_j is judged as malicious because it attempts to attract messages directed towards CID_i which may be dropped.

In *Case I*, FN_j declares a greater group-DP with volunteers of group i, in comparison with $FN_{k,}$. As the claim is supported by the number of encounter tokens signed by volunteers of group i, FN_j is trusted as an appropriate forwarder of messages directed towards CID_i. In *Case II*, FN_j is having a smaller group-DP with volunteers of group i, than $FN_{k,}$ which is warranted by the number of digitally signed encounter tokens it has. But, in *Case III*, FN_j's claim of having a larger group-DP with volunteers of group i in comparison with $FN_{k,}$ (i.e. a claim that it has encountered volunteers of group i more often than $FN_{k,}$) is not supported by the number of digitally signed encounter tokens it has. In reality, $FN_{k,}$, with more such tokens, has clearly met more volunteers of group i than FN_j has. In this case, $FN_{k,}$ identifies FN_j as doubtful, abstains from transmitting messages to it and marks it as a possible black hole. $FN_{k,}$ uses these rules on all other group-DPs and encounter tokens offered by FN_j to make sure that the DPs are untainted.

Owing to the normal development of DPs, an altruistic node will not be eternally banished even if they are momentarily excluded from all sorts of transmission. Forwarder-nodes, who have never met other nodes, will have a less (or no) encounter tokens. However, this would not lead to elimination of these nodes because their group-DP is also not very high. The proposed protocol expels only those nodes that publicize a large group-DP for one or all groups without having adequate number of encounter tokens to prop up the claim.

Therefore, digitally signed encounter tokens offer a genuinely helpful way of detecting black holes and excluding them from future forwarding activities. Hence, SAGE-PRoPHET effectively incorporates security rudiments into the group encounter-based PRoPHET to make certain secured transmission of sensitive situational messages even with existence of unlawful nodes.

5.3.5 Performance Analysis

The effectiveness of SAGE-PRoPHET is evaluated through qualitative analysis and quantitative analysis.

5.3.5.1 Qualitative Analysis

This section presents performance of the proposed SAGE-PRoPHET in terms of storage, computation and energy overheads. The basic assumptions for qualitative analysis remain the same as in Sect. 5.2.4.1.

Storage Overhead

Storage overhead is measured in terms of the memory requirement for storing the different types of pins in the smartphones. In the proposed scheme, every forwarder-node stores n MGPs pertaining to the n groups, one GP pertaining to its own group and S-modified shelter-pins pertaining to the S shelters. So, every forwarder-node maintains $(n + S + 1)$ pins. As 512-bit RSA is used, each pin is of size 512 bits. Thus, memory requirement for storing $(n + S + 1)$ keys in every forwarder-node is $512 * (n + S + 1)$ bits, which is fairly reasonable taking into consideration the storage space of recent smartphones. Also, assuming an average p encounter tokens for each group, a forwarder-node stores, $n \times p$ encounter tokens, for n groups.

Computation Overhead

Each forwarder-node carries out a small number of encryption, decryption and hashing for sending and receiving messages. In the proposed scheme, since forwarder-nodes do not generate keys, computational costs of encryption and decryption only are evaluated. The computational costs, calculated based on the results obtained in [32], are detailed out in Table 5.6.

Table 5.6 Computation and energy overheads of SAGE-PRoPHET

Process	Operation	Algorithm used	Computation overhead [ms]		Energy overhead [mJ]	
Message verification	Decryption	RSA-512	6.46	14.77	7.59	14.45
	Hashing	SHA-256	8.31		6.86	
Receiving message	Hashing	SHA-256	2 × 8.31	24.85	2 × 6.86	23.4
	Encryption	RSA-512	1.77		2.09	
	Decryption	RSA-512	6.46		7.59	
Encounter token generation	Hashing	SHA-256	8.31	10.08	6.86	8.95
	Encryption	RSA-512	1.77		2.09	
Encounter token verification	p Decryption	RSA-512	$p \times 6.46$	$14.77p$	$p \times 7.59$	$14.45p$
	p Hashing	SHA-256	$p \times 8.31$		$p \times 6.86$	
Total			$49.7 + 14.77p$		$46.8 + 14.45p$	

Energy Overhead

The energy costs for the encryption, decryption and hashing occurring at each forwarder-node are computed based on the results obtained in [32]. Message size is assumed to be 1 kB. The results are summarized in Table 5.6.

5.3.5.2 Quantitative Analysis

This subsection provides the quantitative analysis of the proposed SAGE-PRoPHET protocol. The effectiveness of the proposed scheme is evaluated through simulation. The basic assumptions for quantitative analysis of SAGE-PRoPHET remain the same as in Sect. 5.2.4.2.

Simulation Environment

The simulation environment used for quantitative analysis of the multilayer encryption and hashing-based scheme, described in Sect. "Simulation Environment", is used for conducting quantitative performance analysis of SAGE-PRoPHET as well.

Simulation Set-up

The simulation set-up used for the simulation of multilayer encryption and hashing-based scheme, described in Sect. "Simulation Set-up", is used for SAGE-PRoPHET as well with some minor modifications. Along with the nine shelters, five relief camps (corresponding to the five groups) are set up on ONE using the map in [34]. The routing model adopted is PRoPHET [43]. In the simulation, unlawful, malicious and well-behaved nodes are deployed that move around in the affected area. Unlawful forwarders lack a valid GP and MGP pair. Malicious nodes advertise high delivery predictabilities for every node and drop messages. They create false messages and transmit them to nodes they encounter. Altruistic nodes send and receive messages honestly. Table 5.3 enlists the important parameters used in simulation.

Simulation Metrics

The performance metrics used for measuring the effectiveness of the multilayer encryption and hashing-based scheme, described in Sect. "Simulation Metrics", is used for SAGE-PRoPHET as well. In addition to these metrics, two other metrics, defined below, are used to enumerate the efficiency of the proposed protocol.

- ***Average Energy Consumption***: This is used to measure energy efficiency of the proposed protocol. The energy consumption statistics of Samsung Galaxy Grand 2 (with specifications mentioned in Sect. 5.2.4.1) are used to configure the energy module settings of the ONE simulator defined as

$$\text{Average Energy Consumption} = \frac{\text{Total energy consumption in the entire operational period}}{\text{No. of hours in the operational period}}$$

- ***Percentage of Group Messages Delivered***: It is the percentage of messages delivered to a particular camp that are related to that camp. This metric is used to gauge the efficiency of the PEKS-based message categorization technique.

Results and Discussion

The proposed scheme is compared with over other well-known DTN routing protocols like PRoPHET [43], PRoPHETv2 [46], MaxProp [47] and epidemic [48] in terms of the performance metrics defined above. The results presented here are the average of 50 independent runs.

- ***Impact on Delivery Ratio***: Figure 5.6a illustrates the delivery ratio of PRoPHET, PRoPHETv2, MaxProp, epidemic and SAGE-PRoPHET with rising number of malicious nodes. SAGE-PRoPHET defeats all competing schemes with a highest delivery ratio of 87% with 10% unauthorized and malicious nodes. Even though delivery ratio of the proposed protocol drops with rising number of malicious nodes, SAGE-PRoPHET performs better than others.
- ***Impact on Average Delay***: Figure 5.6b demonstrates the average delay of the protocols with rising number of unlawful and malicious nodes. SAGE-PRoPHET has the largest delay due to the computational overhead it faces. This increased

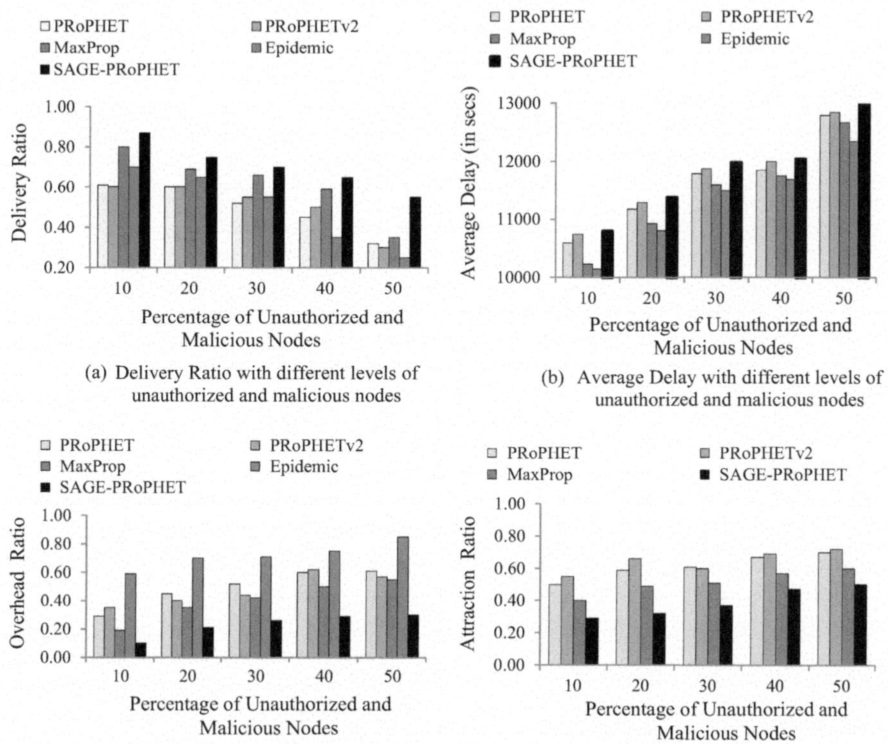

(a) Delivery Ratio with different levels of unauthorized and malicious nodes

(b) Average Delay with different levels of unauthorized and malicious nodes

(c) Overhead Ratio with different levels of unauthorized and malicious nodes

(d) Attraction Ratio with different levels of unauthorized and malicious nodes

Fig. 5.6 Performance of SAGE-PRoPHET with respect to different performance metrics

delay is affordable as it offers increased delivery ratio and reduced attraction ratio. Epidemic performs better than the other competing protocols.

- ***Impact on Overhead Ratio***: Figure 5.6c depicts the overhead ratio of the protocols with increasing number of unauthorized and malicious nodes. SAGE-PRoPHET exhibits 10% overhead ratio in the presence of 10% unauthorized and malicious nodes, which is the least. Epidemic's performance comes out to be worst. Other protocols perform more or less similarly.
- ***Impact on Attraction Ratio***: Figure 5.6d shows the attraction ratio (percentage of messages attracted and dropped by potential black holes) of different protocols with an increasing number of unauthorized and malicious nodes. Here, also SAGE-PRoPHET performs best, with only 29% attraction ratio in the presence of 10% unlawful and malicious nodes, due to the security scheme that detects and avoids potential black holes in the system. MaxProp shows better performance than PRoPHET and PRoPHETv2. Epidemic is not taken into account for this metric because in epidemic a black hole simply drops the messages it receives without

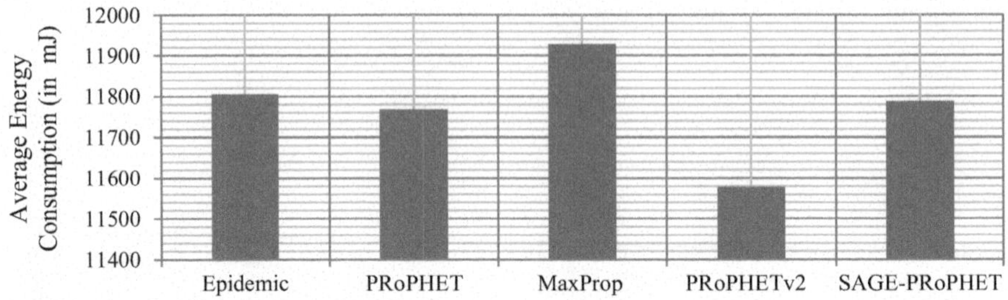

Fig. 5.7 Average energy consumption of different competing protocols

attracting them through fake DP. As a matter of fact, in epidemic routing, a black hole does not have to attract a message, as messages are sent to all nodes.

- ***Average Energy Consumption***: Figure 5.7 presents the average energy consumption of all protocols considering 30% unlawful and malicious nodes. PRoPHETv2 is most energy efficient, whereas MaxProp consumes the maximum energy. Performance of SAGE-PRoPHET is more or less same as PRoPHET. This is due to the fact that the hashing, encryption and decryption operations that SAGE-PRoPHET performs are not that energy consuming, as depicted in Table 5.6. Also, the proposed protocol does not involve extra networking operations like scanning and pairing.
- ***Percentage of Group Messages Delivered***: Figure 5.8 shows the percentage of group and non-group messages in each camp for different protocols. Evidently, SAGE-PRoPHET outperforms all the others and delivers a maximum of 87% group messages to the respective camps.

In this work, a security-aided and group encounter-based PRoPHET routing protocol, called SAGE-PRoPHET, is proposed. First, a PEKS-based message categorization technique is suggested that categorizes messages without compromising on its confidentiality. Next, a group encounter-based PRoPHET routing protocol is suggested that disseminates categorized messages to appropriate relief camps through the best forwarders so that messages reach their destination at the earliest possible. Finally, security features are incorporated into PRoPHET for protecting the messages from eavesdropping attack and network from identity spoofing and bundle store overflow attacks. Thorough security analysis has been conducted. Simulation results depict that the protocol offers 87% delivery ratio, only 10% overhead ratio and 29% attraction ratio with only negligible increase in average delay. Energy consumption of SAGE-PRoPHET is almost similar to PRoPHET and delivers as high as 87% group messages to the respective camps.

Next, a trust-based Watchdog technique is proposed and flawlessly incorporated into PRoPHET to provide security against communication attack like packet drop and trust/reputation attacks like bad-mouthing and ballot-stuffing. The Watchdog-supervised PRoPHET routing protocol ensures successful transmission of crucial

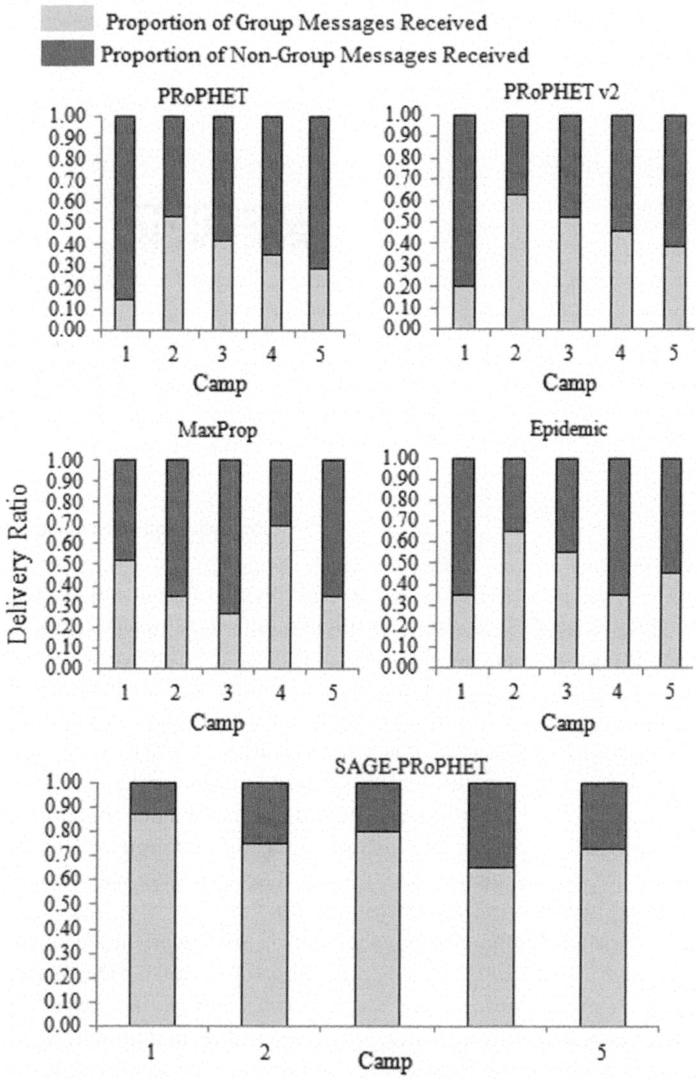

Fig. 5.8 Percentage of group and non-group messages delivered to each camp for different protocols

post-disaster situational information to the control station in the presence of selfish and malicious nodes.

5.4 Wise-PRoPHET: Watchdog-Supervised PRoPHET Routing Protocol

In this section, a trust-based Watchdog technique [49] is presented and seamlessly integrated with PRoPHET to provide security against communication attack like packet drop and trust/reputation attacks like bad-mouthing and ballot-stuffing. The Watchdog-supervised PRoPHET routing protocol ensures successful transmission of crucial post-disaster situational information to the control station even in the presence of selfish and malicious nodes.

Among all other DTN routing schemes, probability-based routings are most frequently used. Such routings, like PRoPHET [43] and MaxProp [47], forward a message to a node with the maximum probability of encountering the destination. Thus, the success of such routing techniques depends on the collaboration from participating nodes, which nevertheless is rather costly for these nodes. Thus, one source of problem in DTNs is the existence of selfish nodes which decline to forward messages for others, to save their own resources. Selfish nodes, by randomly dropping packets, can seriously disrupt transmission of sensitive situational messages to the control station. Another source of problem in DTNs is the existence of malicious nodes that either do bad-mouthing (distributing low trust values for cooperating nodes) or ballot-stuffing (distributing high trust values for selfish nodes) [17]. This disrupts the correct behaviour of the network as a large number of messages are directed towards selfish nodes and are eventually dropped. Thus, identifying and excluding such selfish and malicious nodes while routing messages are important for efficient disaster management. However, the existing PRoPHET routing protocol does not have provisions for addressing these issues.

The Watchdog technique has been widely used to identify selfish and misbehaving nodes in MANETs and WSNs [40, 50–54]. Nevertheless, these techniques either depend on the assumption of the comparatively slow movement of nodes or depend on existing end-to-end routing paths between the source and the destination. None of these Watchdog techniques apply to the underlying smartphone-based DTN, considered in this work, because neither existence of contemporary end-to-end routing paths nor comparatively slow network mobility is practicable in a DTN scenario.

In this work, first, a trust-based Watchdog technique for DTNs is proposed where a Watchdog-enabled node (smartphone used by volunteer/relief worker) observes its neighbouring nodes to create a local perception about their forwarding behaviour—selfish or altruistic. This knowledge is then propagated in the network to build a global perception about the forwarding characteristics of all nodes in the network. The Watchdog is further improved to detect spiteful nodes and prevent the use of incorrect trust information provided by them. Finally, the Watchdog technique is combined with PRoPHET for identifying the ability of a node to deliver messages not only based on its probability of encountering the destination but also depending on its trustworthiness as an honest forwarder. The blending of Watchdog scheme and PRoPHET is done in a manner that rationalizes self-trusting, a property of trust-based data forwarding in opportunistic networks which reduces and delays message transfers. This scheme enables further improvement of delivery ratio and delay. The proposed Wise-PRoPHET scheme successfully identifies and avoids selfish and malicious nodes and routes' vulnerable post-disaster situational messages through the best possible forwarders.

5.4.1 System Model

In this section, the system model related to proposed protocol is described. The system model includes network architecture, routing protocol, threat and trust models and id generation.

5.4.1.1 Network Architecture

The underlying network architecture of the proposed protocol is mostly the same as the network architecture used for the multilayer encryption and hashing-based scheme, illustrated in Sect. 5.2.1.1.

5.4.1.2 Routing Protocol

The forwarder-nodes share the messages with others following the DTN routing protocol PRoPHET [43]. Details of the PRoPHET routing protocol are already discussed in Sect. 5.3.1.3.

5.4.1.3 Attack Model

Even though there can be various kinds of misbehaving nodes that may create serious security threats to a DTN, only selfish and malicious nodes are considered in this work. A selfish node uses the routing services through other nodes but does not contribute its own resources to cooperate towards that service [17]. Malicious nodes are a group of nodes that conspire to launch attacks like bad-mouthing (ruining the reputation of altruistic nodes) and ballot-stuffing (boosting the reputation of selfish nodes). The attacks launched by these nodes are discussed below:

- ***Packet Dropping Attack***: In this attack, a selfish node chooses to drop packets entirely or selectively.

- ***Bad-mouthing Attack***: This is ruining the reputation of well-behaved nodes. Malicious nodes may set the trust value of all altruistic nodes to 0.
- ***Ballot-stuffing Attack***: This is boosting the reputation of selfish nodes where malicious nodes set the trust value of all selfish nodes to 1.

5.4.1.4 Trust Model

The trust level of a forwarder-node is defined as a real number in the range of [0, 1], with 1 indicating full trust (fully unselfish), 0.5 ignorance and 0 complete mistrust (fully selfish). This work deals with two types of trusts as described below:

- ***Direct Trust***: *Direct trust* of a forwarder-node is the information about the node's forwarding behaviour provided by the Watchdog that directly monitors the node. Direct trust is subjective (opinion about a node's trustworthiness varies from node to node) and asymmetric (two nodes may not have similar mutual trust).
- ***Indirect Trust***: Direct trust value of a specific node is shared with other nodes in the network. Other nodes use this second-hand information as indirect trust about that particular node. Such indirect trusts facilitate faster diffusion of trust information across the network and are used to calculate four types of trust values. Local trust of a forwarder-node is the average of direct trust values of that node, supplied by a group of nodes. It depicts a local perception about the forwarding behaviour of that node. Global trust of a forwarder-node is the average of direct trust values of that node, provided by all other nodes in the network, providing global perception about the forwarding behaviour of that node. Forwarder trust of a forwarder-node is the weighted average of direct and global trusts. Rater trust of a forwarder-node is the conformance of the direct trust values it provides (about other nodes) with the average direct trust value of those nodes.

5.4.1.5 Node- and Message-id Generation

Each node (shelter-nodes, forwarder-nodes and control-node) and each message are assigned a unique id in the proposed Watchdog technique. However, synchronizing these ids in a DTN environment becomes challenging as no end-to-end communication exists among the nodes.

In the proposed technique, as in Sect. 5.3.1.5, the central control station, in the set-up phase, assigns a shelter-id, say SID_s, $s = 1, 2, \ldots, S$, to each shelter-node, a forwarder-id, say FN_i, $i = 1, 2, \ldots, N$, to each forwarder-node and a control-id, say CN, to the control-node. When a new shelter is established, the corresponding shelter-node has to register with the control station to obtain a shelter-id. Similarly, a new volunteer has to register with the control station to get a forwarder-id if she wants to join the relief operation. Thus, new entrants can be always included into the network. Apart from generation of these ids, the central control station has no role in the run-time phase. Forwarder-nodes exchange situational messages using the

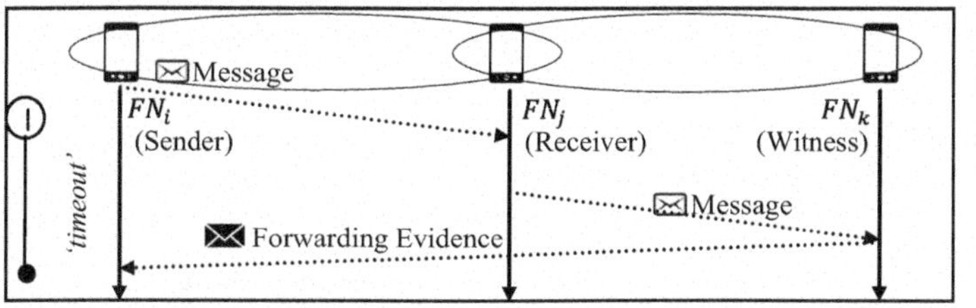

Fig. 5.9 Forwarding evidence mechanism

Wise-PRoPHET technique during run-time phase. A message-id is created for each message at the time of its generation. Message-id is essentially a unique message number prefixed with the source-id, the id of the node (may be shelter-node or forwarder-node or control-node) that generates the message.

5.4.2 Proposed Wise-PRoPHET Technique

This section proposes the Wise-PRoPHET technique where a node chooses the next-hop forwarder based on the forwarder's probability of encountering the destination and its trustworthiness as an honest forwarder. The technique uses three modules—(i) Watchdog operation, (ii) indirect trust enumeration and (iii) forwarder selection.

5.4.2.1 Watchdog Operation

Watchdog operation module consists of two phases. In the first phase, a node, with the installed Watchdog, monitors the forwarding behaviour of its neighbouring nodes. In the second phase, that node, based on the monitored behaviour, calculates direct trust values of the neighbouring nodes.

Monitoring Forwarding Behaviour

When forwarder-node FN_i meets another node FN_j, FN_i monitors FN_j on the basis of a forwarding evidence (*FE*) sent by FN_k. FN_k is the node to which FN_j has forwarded the message. The function of the *FE* is to convince FN_i that FN_j has actually forwarded the message to a competent forwarder FN_k. Figure 5.9 depicts the mechanism. The *FE* creation, circulation and quantification processes are explained below. The presented technique partially adopts the *FE* creation and propagation process presented in [1]. Nevertheless, the current technique of *FE* quantification is different from [1]; instead of classifying them as only "good" and "bad", this technique objectively quantifies the evidences for computing direct trust values.

Message ID	Source ID	Destination ID	Generation Time	Expiration Time	Sender ID	Receiver ID	Witness ID	Receiver DP	Witness DP	Sending Time	Receiving Time
Fixed Attributes					Variable Attributes						

Fig. 5.10 Twelve-tuple *FE* structure

- ***FE Creation***: To begin with, when FN_i encounters FN_j at time t, FN_i checks the delivery predictability of FN_j (refer to Sect. 5.3.1.3) and transmits a message to it. After transmitting the message, FN_i waits for a *timeout* period for the arrival of the forwarding evidence $FE_{i,j}(t)$ related to the message. *timeout* is the expected time required for the evidence to reach FN_i. A rational *timeout* period is derived in Sect. 5.4.3.3. FN_j carries the message until it encounters a more competent node, FN_k, and transmits it to FN_k. On receiving the message, FN_k generates $FE_{i,j}(t)$ directed to FN_i. $FE_{i,j}(t)$ has 12 attributes; the 12-tuple *FE* structure is shown in Fig. 5.10.
- ***FE Circulation***: The $FE_{i,j}(t)$ generated by FN_k, directed to FN_i, is transmitted over the network using epidemic routing [48]. $FE_{i,j}(t)$ is required to be fast percolated in the network so that it quickly reaches FN_i in multiple hops. Thus, it becomes imperative to exploit the high delivery ratio property of a flooding-based routing like epidemic. Since the size of the *FE*s is quite less (12×4 bytes $= 48$ bytes), the overhead introduced by epidemic routing does not degrade the network performance substantially. Moreover, FN_k can club together multiple *FE*s directed to FN_i, which further lessens the overhead. On receiving $FE_{i,j}(t)$, all other nodes except FN_i keep on forwarding it until the message expires or the evidence reaches FN_i. FN_i on receiving $FE_{i,j}(t)$ in multiple hops stops forwarding it further. If FN_k gets an opportunity to encounter FN_i, it delivers the evidence to FN_i and stops further retransmission. Unlike the end-to-end acknowledgements used in MANETs, *FE*s do not rely on contemporary end-to-end routing paths between the source and the destination and get transmitted to FN_i in single or multiple hops.
- ***FE Quantification***: On receiving $FE_{i,j}(t)$, all other nodes except FN_i transmit it, while FN_i stops forwarding it further. Other nodes also stop forwarding the evidence when the message expires. FN_i verifies the goodness of the forwarding behaviour of FN_j in terms of forwarding score, $FS_{i,j}(t)$, computed as follows:

$$FS_{i,j}(t) = \begin{cases} 1 \text{ if } WitnessDP \geq ReceiverDP \\ 0 \text{ otherwise} \end{cases}$$

Thus, FN_j is given full credit only if it transmits the message to a more competent node. Thus, forwarding score is assigned after examining the quality of forwarding. It is to be noted that—(i) FN_j transmits the message received from FN_i to multiple FN_ks using PRoPHET, and (ii) each FN_k percolates $FE_{i,j}(t)$ through multiple forwarder-nodes using epidemic. Owing to the above facts, FN_i may receive multiple $FE_{i,j}(t)$ related to a single message transmitted to FN_j. Nevertheless, FN_i considers only the

Table 5.7 Trust table stored at FN_i

Node-id		FN_1	…	FN_j	…	FN_N
Direct trust		$T_{i,1}^{direct}(t)$	…	$T_{i,j}^{direct}(t)$	…	$T_{i,N}^{direct}(t)$
Indirect trust	Local trust	$T_{i,1}^{local}(t)$	…	$T_{i,j}^{local}(t)$	…	$T_{i,N}^{local}(t)$
	Global trust	$\hat{T}_{i,1}^{global}$	…	$\hat{T}_{i,j}^{global}$	…	$\hat{T}_{i,N}^{globaal}$
	Forwarder trust	$T_{i,1}^{forwarder}$	…	$T_{i,j}^{forwarder}$	…	$T_{i,N}^{forwarder}$
	Rater trust	$T_{i,1}^{rater}$	…	$T_{i,j}^{rater}$	…	$T_{i,N}^{rater}$

$FE_{i,j}(t)$ related to a message that reaches it first. FN_i also increments a selfishness score, $SS_{i,j}(t)$, for FN_j as follows:

$$SS_{i,j}(t) = \begin{cases} \text{incremented} & \text{if } FN_i \text{ does not receive any } FE_{i,j}(t) \text{ within '}timeout\text{'} \\ \text{not incremented} & \text{otherwise} \end{cases}$$

Computing Direct Trust

Forwarding scores (FS) and selfishness score (SS) offered by the Watchdog at the monitoring phase are used to assign direct trust value to the monitored node. For example, FN_i assigns direct trust value $T_{i,j}^{direct}(t)$ to FN_j based on all received FSs and SSs about FN_j based on all messages forwarded to FN_j at time t. Such direct trust about FN_j assigned by FN_i is computed as:

$$T_{i,j}^{direct}(t) = \begin{cases} 0 & \text{if } \sum_{AllMsg} FS_{i,j}(t) = 0 \\ 0 & \text{if } \sum_{All\,Msg} SS_{i,j}(t) > \sum_{All\,Msg} FS_{i,j}(t) \\ 1 - \frac{\sum_{All\,Msg} SS_{i,j}(t)}{\sum_{All\,Msg} FS_{i,j}(t)} & \text{otherwise} \end{cases} \quad (5.1a)$$

Each node in the network computes direct trust about all other nodes in the network using Eq. 5.1a and stores them in the trust table, shown in Table 5.7. Direct trust value of nodes that have not been monitored is set to 0.5.

Updating Direct Trust

FN_i updates direct trust about FN_j according to Eq. 5.1b shown below:

$$T_{i,j}^{\prime\,direct}(t+\Delta t) = \begin{cases} \mu\, T_{i,j}^{direct}(t) + (1-\mu) T_{i,j}^{direct}(t+\Delta t) & \text{if } FN_i \text{ can monitor } FN_j \text{ in } \Delta t \\ e^{-\theta \Delta t}\, T_{i,j}^{direct}(t) & \text{otherwise} \end{cases} \quad (5.1b)$$

Here, $T_{i,j}^{direct}(t+\Delta t)$ is the direct trust about FN_j at time point $(t+\Delta t)$, assigned by the Watchdog. $T_{i,j}^{\prime direct}(t+\Delta t)$ is the updated direct trust about FN_j at time point $(t+\Delta t)$. Since forwarding behaviour of nodes keeps on changing with time between selfish and altruistic, a weighing factor $\mu, 0 \leq \mu \leq 1$, is used to balance previous and current trust enumerations.

Considering the fact that trustworthiness degrades with time, direct trust value is multiplied with a decaying factor $e^{-\theta \Delta t}$, with $0 < \theta \leq 0.1$, if FN_i does not get a chance to encounter FN_j. These updated direct trust values are disseminated in the network so that nodes can use them in the indirect trust enumeration phase.

5.4.2.2 Indirect Trust Enumeration

This module contains four steps. In the first step, a node collects direct trusts about other nodes in the network from its neighbours and uses these direct trusts to compute local trusts. In the second step, the node, using the local trusts, estimates global trusts of all other nodes in the network. As the third step, the node, based on these global trusts, computes forwarder trusts for identifying selfish nodes. In the last step, the collected direct trusts are further used to calculate rater trusts for identifying malicious nodes.

Computing Local Trusts

When FN_i comes in contact with a group of nodes at particular time, it collects direct trust values (about all other nodes in the network) from these nodes. Suppose FN_i encounters n nodes, say $FN_1, FN_2, \ldots, FN_n$, at contact time t. FN_i collects direct trust values, $(T_{1,j}^{direct}, T_{2,j}^{direct}, \ldots, T_{n,j}^{direct})$, about $FN_j, j = 1, 2, \ldots, N, j \neq i$, from these nodes and uses them to compute a local trust value $T_{i,j}^{local}(t)$ of FN_j as:

$$T_{i,j}^{local}(t) = \frac{1}{n} \sum_{k=1}^{n} T_{k,j}^{direct} \tag{5.2a}$$

FN_i stores these local trust values for all other nodes, FN_j, in the network in the trust table, shown in Table 5.7. Local trust values of nodes for which there is no recommendation yet are set to 0.5.

Estimating Global Trusts

The direct trust values about a specific node received from various sources may vary to a great extent, and there is hardly any global consensus. Thus, it is logical to consider such direct trust values from all nodes in the network to get the global perception about the specific node. In particular, the global trust of FN_j maintained at FN_i is defined as the average of direct trust values about FN_j offered by all $(N - 2)$ nodes in the network. Global trust can be represented as:

$$T_{i,j}^{global} = \frac{1}{N-2} \sum_{\substack{k=1 \\ k \neq i,j}}^{N} T_{k,j}^{direct} \tag{5.2b}$$

The global trust helps a node build a more accurate picture of other nodes' forwarding behaviour. Nevertheless, in a DTN environment, a node comes in contact with a limited number of nodes at a given time and hence collects a few sample direct trust values about a specific node. Thus, it is not practicable to collect the entire population of direct trust values and enumerate the global trust based on those. Statistical estimation [55] is used to estimate the global trust of a node from a set of sample direct trust values for that node.

The value of direct trust (lying between 0 and 1) of a forwarder-node is determined by the events—(i) the forwarder-node transmits a message received from a previous-hop to the next-hop, and (ii) a forwarding evidence from the next-hop is received by the previous-hop. Both these events are highly stochastic in nature. Thus, direct trust can be considered as a continuous random variable and is supposed to follow a uniform distribution over the interval [0, 1] with mean 0.5. The probability density function of direct trust is given by the formula

$$f = \begin{cases} \frac{1}{1-0} = 1, \; 0 \leq \textit{direct trust} \leq 1 \\ 0 \qquad\qquad \text{otherwise} \end{cases}$$

For estimating global trust, from a set of sample direct trusts, using statistical estimation, the following are considered:

Total number of nodes: N
Population size: $N - 2$, excluding FN_i and FN_j
Population: $\{T_{1,j}^{direct}\}, i = 1, 2, \ldots, N-2$
Population mean: $T_{i,j}^{global}$, global trust of FN_j to be stored at FN_i
Sample size: n
Sampling scheme: Simple random sampling without replacement
Probability of a sample: $1/^N C_n$
Sample: $\{T_{i,j}^{direct}\}, i = 1, 2, \ldots, n,$
Sample mean: $T_{i,j}^{local}$, to be stored at FN_i

Following the theory of random sampling, $T_{i,j}^{local}$ can be considered as a random variable that assumes different values for different samples with probability $1/^N C_n$. In the presented technique, FN_i repeats the simple random sampling S times at consecutive time points $(t, t + \Delta t, \ldots, t + (S-1)\Delta t)$ and computes local trust values of FN_j as $\{T_{i,j}^{local}(t), T_{i,j}^{local}(t + \Delta t), \ldots, T_{i,j}^{local}(t + (S-1)\Delta t)\}$. These local trusts, i.e. the different values of the sample mean $T_{i,j}^{local}$, each occurring with probability $1/^N C_n$, are stored in a sampling distribution of local trust, shown in Table 5.8.

Now according to theory of statistical estimation, sample mean is an unbiased estimator of the population mean and expectation of sample mean is an unbiased

Table 5.8 Sampling distribution of local trust values of FN_j

$T_{i,j}^{local}$	$T_{i,j}^{local}(t)$	...	$T_{i,j}^{local}(t+(S-1)\Delta t)$
Probability	$1/^N C_n$	...	$1/^N C_n$

estimate of the population mean. In the present context, local trust, $T_{i,j}^{local}$, is an unbiased estimator of global trust, $T_{i,j}^{global}$ and

$$\hat{T}_{i,j}^{global} = E\left(T_{i,j}^{local}\right) = T_{i,j}^{local}(t) \times 1/^N C_n + \ldots + T_{i,j}^{local}(t+(S-1)\Delta t) \times 1/^N C_n \tag{5.2c}$$

The accuracy of the estimated global trust depends upon the number of samples (S, in this case), i.e. the number of local trusts, being considered for estimation. FN_i similarly estimates the global trust values of all other nodes in the network and stores them in the network in the trust table, shown in Table 5.7. Such estimated global trust is then combined with the direct trust to obtain a comprehensive forwarder trust value for a node.

Combining Direct and Global Trusts to Form Forwarder Trust

To judge the forwarding behaviour of FN_j, FN_i combines the direct and global trusts FN_j to form a forwarder trust value as

$$T_{i,j}^{forwarder} = \omega_1 T_{i,j}^{direct} + \omega_2 \hat{T}_{i,j}^{global} \tag{5.2d}$$

where $T_{i,j}^{direct}$ and $T_{i,j}^{global}$ are the last updated direct trust and global trust of FN_j maintained by FN_i. The weighing factors ω_1 and ω_2 $(0 \le \omega_1, \omega_2 \le 1, \omega_1 + \omega_2 = 1)$ are used to decide the importance of direct observation against recommendations. FN_i stores these direct trust values for all other nodes in the network in the Trust Table, shown in Table 5.7. These forwarder trust values are used for enumerating the forwarding competency of a prospective forwarder which is explained in Sect. 5.4.2.3.

Computing Rater Trust

To alleviate bad-mouthing and ballot-stuffing, FN_i judges the correctness of the direct trust values offered by FN_j by computing its rater trust. To compute rater trust, FN_i calculates the deviations of direct trust values shared by FN_j, about other nodes, from the average direct trust value (i.e. local trust value) of these nodes. These deviations indicate to what extent recommendations offered by FN_j diverge from the recommendations offered by others. FN_i finds the mean of deviations (MD) at time t as below:

$$MD_{i,j}^{direct}(t) = \frac{1}{N-2} \sum_{\substack{l=1 \\ l \neq i,j}}^{N} \left| T_{j,l}^{direct}(t) - T_{i,l}^{local}(t) \right| \tag{5.2e}$$

To bring rationality to the judgment process, FN_i computes the mean of deviations for S consecutive time points. The rater trust value is then computed as follows:

$$T_{i,j}^{rater} = 1 - \frac{1}{S} \sum_{m=0}^{S-1} MD_{i,k}^{direct}(t + m\Delta t) \tag{5.2f}$$

If value of $T_{i,j}^{rater}$ falls below a specified threshold τ (set through simulation), FN_i considers FN_j as malicious and refrains from using the direct trust values it provides. FN_i stores rater trust values for all other nodes in the trust table, shown in Table 5.7.

5.4.2.3 Forwarder Selection

In this module, a node first enumerates the forwarding competency of a prospective forwarder and uses this competency to select the best next-hop forwarder. Forwarder trust of a node determines its forwarding behaviour (selfish or altruistic). Its delivery predictability, on the other hand, depicts the probability of encountering the final destination of the message. Thus, a node while taking forwarding decision integrates these metrics to find forwarding competency (FC) of the forwarder as shown below

$$FC_j = T_{i,j}^{forwarder} \times P\left(FN_j, DestID\right) \tag{5.3}$$

Whenever FN_i encounters FN_j and intends to transmit a message, FN_i computes FC_j using Eq. (5.3) and takes forwarding decision according to the following rule:

$$\begin{array}{ll} \text{Forwards message to } FN_j & \text{if } FC_j > FC_i \\ \text{Don't forward message to } FN_j, & \text{otherwise} \end{array}$$

The rationale behind this multiplicative model in Eq. (5.3) is that if FN_j is selfish, FN_i assigns a low forwarder trust to FN_j and $FC_{i,j}$ is decreased. However, this model suffers from *self-trusting*, a property of trust-based data forwarding in opportunistic networks. Self-trusting refers to the fact that nodes trust themselves more than others [18], implying that FN_i sets $T_{i,i}^{forwarder}$ to 1 in all situations. Thus, the forwarding competency of FN_i to itself, i.e. $FC_{i,i}$, is only equal to its delivery predictability $P(FN_i, DestID)$. As a result, FN_i attempts to directly deliver messages to the destination itself, unless it encounters another node with much stronger delivery predictability and/or extremely high forwarder trust. The effect of self-trusting is increased delay in delivering messages to the destination, as most of the nodes get

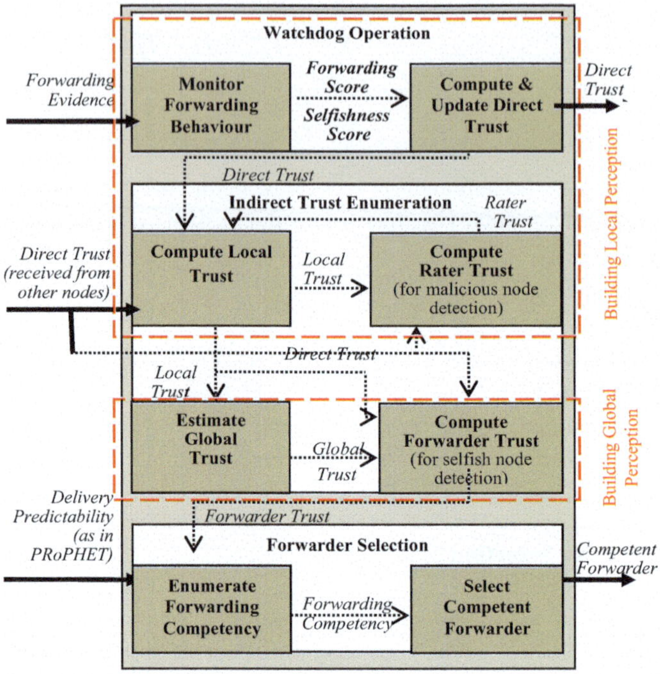

Fig. 5.11 Proposed Wise-PRoPHET framework

rejected as competent forwarders. This is demonstrated in the quantitative performance analysis section.

To combat the effect of self-trusting, a self-trust rationalizing factor, say α, is introduced. Whenever FN_i encounters FN_j and intends to forward a message, FN_i computes $FC_{i,j}$ using Eq. (5.3) and computes $FC_{i,i}$ as

$$FC_{i,i} = (\alpha \times T_{i,i}^{forwarder}) \times P(FN_i, DestID) \tag{5.4}$$

where α ranges from 0.5 to 1. Assuming α as 0.5 will allow FN_i to select forwarders with moderate delivery predictability and forwarder trust. This results in faster delivery of more number of messages, although at the risk of incorporating some selfish nodes into the forwarding activity. Taking α as 1 compels FN_i to choose forwarders with high delivery predictability and/or high forwarder trust. This results in slower delivery of lesser number of messages, ensuring the elimination of selfish nodes from the forwarding activity.

The three interoperable modules of Wise-PRoPHET work together for identifying selfish and malicious nodes in the network and forwarding post-disaster situational messages through the best possible forwarders. Figure 5.11 elucidates the technique.

5.4.3 The Wise-PRoPHET Algorithm

In this section, an algorithm for the proposed Wise-PRoPHET technique is presented.

Input: Forwarding Evidence, Direct Trust, Delivery Predictability
Output: Direct Trust, Competent Forwarder
Runs At: Forwarder-node (FN_i, DTN enabled smartphone, carried by volunteers and relief workers)

```
// WATCHDOG OPERATION BEGINS
    if Rater Trust of FN_j > τ then
       accept direct trust values from FN_j
    endif
    share direct trust values with FN_j
    for MID = 1 to M do // M Messages
// Direct Trust Enumeration
          if FE for MID received within 'timeout' then
             if WitnessDP ≥ ReceiverDP then
                Forwarding Score = 1
             else
                 Forwarding Score = 0
             endif
          else
             Increment Selfishness Score
          endif
    end for
    Compute Direct Trust of FN_j //using Eq. 6.1(a)
    Update Direct Trust of FN_j //using Eq. 6.1(b)
// WATCHDOG OPERATION ENDS
// INDIRECT TRUST ENUMERATION BEGINS
// Local Trust Computation
       Compute Local Trust of FN_j //using Eq. 6.2(a)
// Global Trust Estimation
     Estimate Global Trust of FN_j //using Eq. 6.2(c)
// Forwarder Trust Computation
    Compute Forwarder Trustof FN_j //using Eq. 6.2(d)
// Rater Trust Computation
    for m = 0 to S − 1 do // assuming S timepoints
       Compute Mean of Deviations //using Eq. 6.2(e)
    endfor
    Compute Rater Trust of FN_j //using Eq. 6.2(f)
 end for
// INDIRECT TRUST ENUMERATION ENDS
// FORWARDER SELECTION BEGINS
 for MID = 1 to M do // M Messages
   Compute Forwarding Competency of FN_j using Eq. 6.4
   if FC_j > FC_i then
       transfer MID to FN_j
   endif
 end for
// FORWARDER SELECTION ENDS
```

5.4.3.1 Algorithmic Complexity

In the worst case, the Watchdog operation module runs for a maximum of M times for all messages (M). So, the complexity is $O(M)$. In the indirect trust enumeration module, local trust computation runs S times corresponding to the S time points. Similarly, rater trust computation also runs for S times. Thus, the complexity can be derived as $O(S+S) = O(2S) = O(S)$.

Each of Watchdog operation, indirect trust enumeration and forwarder selection modules run for a maximum $(N-1)$ time, as a forwarder-node FN_i can encounter at most $(N-1)$ other nodes in the network. Thus, the worst-case complexity of the algorithm is

$$O(N(M+S))$$

5.4.3.2 Correctness of the Algorithm

The trust-based forwarding competency of a node is used to prove the correctness of Wise-PRoPHET algorithm.

Lemma The forwarding competency of a next-hop forwarder determines whether a message gets forwarded to it.

Proof Let A be the event that the next-hop forwarder of a node is not selfish. Then, A^c is the event that it is selfish. Also, let B be the event that the message is forwarded to the next-hop forwarder. Since, A and A^c are mutually exclusive and exhaustive,

$$B = (A \cap B) \cup \left(A^c \cap B\right)$$

Since the events $(A \cap B)$ and $(A^c \cap B)$ are also mutually exclusive, applying the theorem of total probability,

$$P(B) = P(A \cap B) + P\left(A^c \cap B\right)$$

Applying theorem of compound probability,

$$P(B) = \{P(A) \times P(B/A)\} + \left\{P\left(A^c\right) \times P\left(B/A^c\right)\right\} \tag{5.5}$$

The probability that the message gets forwarded to the next-hop forwarder, given that it is not selfish, is determined by the delivery predictability of the forwarder, $P(\textit{next-hop, DestID})$. Thus,

$$P(B/A) = P(\textit{next-hop}, \textit{DestID})$$

Also, probability that the message gets forwarded to the next-hop forwarder, given that it is selfish, is 0, i.e.

$$P(B/A^c) = 0$$

Also, $P(A) = T_{next\text{-}hop}^{forwarder}$ and $P(A^c) = 1 - T_{next\text{-}hop}^{forwarder}$. Putting these values in Eq. (5.4),

$$P(B) = T_{next\text{-}hop}^{forwarder} \times P(next\text{-}hop, DestID)$$

which is in fact the forwarding competency of the next-hop forwarder derived in Eq. (5.3). Thus, the trust-based forwarding competency of a next-hop forwarder determines whether a message gets forwarded to it. This proves the correctness of the technique.

5.4.3.3 Deriving a Rational "*timeout*" Period

As explained in Sect. "Monitoring Forwarding Behaviour", FN_i after transmitting a message to FN_j at time t waits for a *timeout* period for the arrival of a forwarding evidence $FE_{i,j}(t)$ related to the message. Return (or non-return) of the forwarding evidence within the *timeout* period has a major impact on the direct trust about FN_j, as explained in Eq. 5.1a. Thus, it becomes imperative to derive a rational *timeout* period for which FN_i must wait before incrementing the selfishness score of FN_j.

As illustrated in Fig. 5.10, *timeout* should be the total time required for the message to reach FN_k from FN_j (using the Wise-PRoPHET technique) and the forwarding evidence to reach FN_i from FN_k (using epidemic routing). So, it is attempted to compute the expected message delay first for epidemic routing and then for the presented Wise-PRoPHET technique. Transmissions between two nodes are supposed to take place at encountering times of the nodes and instantaneous; i.e. the transmission time of a message is very less with respect to the encountering times. Hence, message delay refers to the time required for a node to find the next possible forwarder and deliver the message to that forwarder. To compute the expected message delays, a stochastic model is introduced that models message delay between any two nodes in a DTN. The works presented in [56] and [57] are adapted to derive expressions for expected message delay in the presented technique.

Consider the network with N identical forwarder-nodes, as stated in Sect. 5.2.1.1. Suppose there is a single message to be delivered by a source node to a destination node. Intermediary nodes are used as relay nodes. Let $0 \le t_{p,q}(1) < t_{p,q}(2) < \cdots$ be the successive encountering times between forwarder-nodes FN_p and FN_q. Define $\delta_{p,q} = t_{p,q}(n+1) - t_{p,q}(n)$, as the nth inter-contact time between FN_p and FN_q. The processes, $\{t_{p.q}(n), n \ge 1\}, 1 \le p, q \le N, p \ne q$, are mutually independent and identically distributed Poisson processes with rate $\lambda > 0$ [57]. Equivalently, the random variables $\{\delta_{p,q}\}$ are mutually independent and exponentially distributed

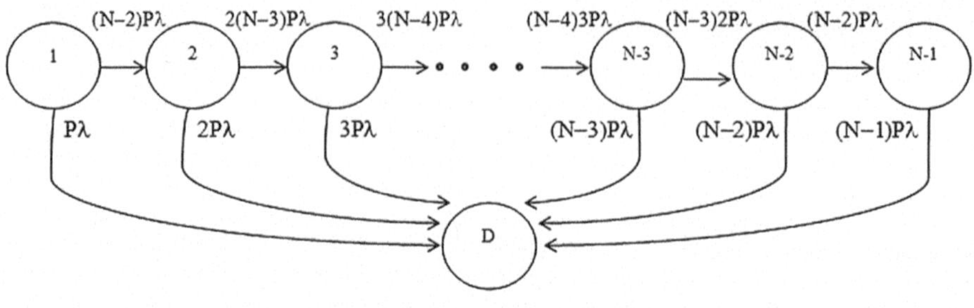

Fig. 5.12 Transition diagram of the Markov chain for multicopy routing

with mean $1/\lambda$, and $\lambda > 0$ is the parameter of the distribution, often called the rate parameter and $1/\lambda$ average inter-contact length. The number of copies of the message in the network can be modelled as an absorbing finite-state Markov chain. The Markov chain takes its value in $\{1, 2, \ldots, N\}$. The Markov chain is in state $i = 1, 2, \ldots, (N-1)$ when there are i copies of the message in the network including the original message, and it is in state N when the message is delivered to the destination node. The states $1, 2, \ldots, (N-1)$ are transient states, whereas N is an absorbing state.

The transition diagram of the Markov chain for a multicopy routing protocol (controlled or uncontrolled) is shown in Fig. 5.12. In this protocol, each node which has a copy of the message forwards it to a node that does not have a copy and which comes within its transmission range. The message is forwarded to an encountered node with a probability P if it is not the destination and with a probability 1 if it is the destination. Thus, when there are i copies of the message in the network, a new copy is generated at the rate $iP\lambda(N-1-i)$ (transition from state i to $i+1$) and one of those i copies reaches the destination node at the rate $i\lambda$ (transition from state i to N), as illustrated in Fig. 5.12. The chain jumps from state i to $i+1$ with probability $P(N-i-1)/(P(N-i-1)+1)$, and it jumps from state i to N with probability $1/(P(N-i-1)+1)$.

In Wise-PRoPHET (controlled multicopy routing), a node forwards the message to its next-hop forwarder with a probability P, where

$$P = T_{\textit{next-hop}}^{\textit{forwarder}} \times P(\textit{next-hop}, \textit{DestID})$$

if the next-hop forwarder is not the destination and with a probability 1 if it is the destination. Using results from [57], the expected message delay (*EMD*) for Wise-PRoPHET is derived as:

$$EMD_{\textit{Wise-PRoPHET}} = \frac{1}{\lambda(N-1)} \sum_{i=1}^{N-1} \sum_{j=1}^{i} \frac{1}{PNj - Pj^2 + Pj + j} \tag{5.6a}$$

which is the time taken by the message to reach FN_k from FN_j. P, lying between 0 and 1, differs for each next-hop forwarder.

In epidemic (uncontrolled multicopy) routing, a node forwards the message to its next-hop forwarder with a probability 1 irrespective of whether it is the destination or not, i.e. for such a routing $P = 1$. Using results from [56], the expected message delay (*EMD*) for epidemic routing is derived as:

$$EMD_{epidemic} = \frac{1}{\lambda(N-1)} \sum_{i=1}^{N-1} \frac{1}{i} \tag{5.6b}$$

which is the time taken by the forwarding evidence to reach FN_i from $\mathrm{FN_k}$. Thus, a rational value for *timeout* can be obtained by combining the results derived in Eqs. 5.4a and 5.4b as

$$timeout = EMD_{Wise\text{-}PRoPHET} + EMD_{epidemic} \tag{5.6c}$$

It can be observed from Eq. 5.4(a) that the minimum value of $EMD_{Wise\text{-}PRoPHET}$ is $\frac{1}{\lambda(N-1)} \sum_{i=1}^{N-1} \frac{1}{i}$, obtained when $P = 1$. Thus, using Eq. 5.4c, the minimum value of *timeout* (i.e. the minimum time for which FN_i must wait before considering FN_j as selfish) is obtained as $\frac{2}{\lambda(N-1)} \sum_{i=1}^{N-1} \frac{1}{i}$. For practical values of $1/\lambda = 1800$, 3600, 7200 s [58] and $N = 100$, the minimum value of *timeout* comes out to be 180, 360 and 720 s, respectively. These *timeout* values are used for quantitative performance evaluation of the presented technique in Sect. 5.4.4.2.

5.4.4 *Performance Analysis*

The efficiency of the Wise-PRoPHET technique is measured by both qualitative and quantitative analyses.

5.4.4.1 Qualitative Analysis

This section provides overhead analysis of the Wise-PRoPHET technique which is run on resource-limited devices like smartphones (forwarder-node). The worst-case computation and communication overheads are computed for an operational period, which is typically 12 h for the present application. The worst-case scenario arises when the node encounters all other $(N - 1)$ nodes in the network. It is supposed that a node on an average has P contacts with every other node, encounters n nodes at time t and exchanges M messages. Storage overhead is not measured considering the substantial storage facilities in smartphones. The smartphone specifications used for the qualitative analysis remain the same as in Sect. 5.2.4.1.

Computation Overhead

The computation overhead of a forwarder-node for executing Wise-PRoPHET is the total number of low-end operations [59] it performs. Table 5.9 depicts the required number of such low-end operations and corresponding energy requirements. Using these, the computation energy overhead sums up to:

$$3.5MN + 0.9Mn + 2.3NS + 0.2n^2 - 4.4M + 1.5N - 2.1S + 0.2n - 1.8\,\mathrm{J}$$

Communication Overhead

Communication overhead is measured in terms of the total number of bytes exchanged by a forwarder-node for executing the Wise-PRoPHET technique. As per the assumption stated above, a node can encounter all other $(N - 1)$ nodes in the network with P contacts per node. On each such contact, the forwarder-node exchanges direct trust values of $(N - 1)$ other nodes. Thus, the node transmits and also receives $((N - 1)(N - 2)P)$ direct trust values. Assuming M transmitted and Q received forwarding evidences in the entire operational period, a forwarder-node transmits $12M$ and receives $12Q$ values in the entire operational period. Hence, assuming k bytes for each transmitted/received value, communication overhead sums up to:

$$\begin{aligned} \textit{Transmission Overhead}: &\ k\left(N^2P - 3NP + 12M + 2P\right)\ \text{bytes} \\ \textit{Reception Overhead}: &\ k\left(N^2P - 3NP + 12Q + 2P\right)\ \text{bytes} \end{aligned}$$

The Bluetooth communication energy consumption for the smartphone under consideration is 0.384 J/s for transmitting and 0.329 J/s for receiving 1 Mb data [31]. Also, Bluetooth 1.2 has an average data transfer rate of 1 Mb/s. Thus, the communication energy overhead is computed as

Table 5.9 Operations and corresponding energy consumptions

Operation	Number of operations	Energy consumption (J)/operation
LOAD	$7MN + 6NS - 7M + 2N - 6S - 2$	0.23
MOV	$4MN + NS - 4M + 4N - S - 4$	0.20
STORE	$3MN \pm 3M + 2N - 2$	0.12
CMP	$MN + NS - M - S$	0.75
MUL	$5Mn + n^2 - 5M + n + S - 2$	0.18

$$\textit{Transmission Energy Overhead}: \ 0.384/(1024)^{-2} \times k\left(N^2P - 3NP + 12M + 2P\right) \text{ J}$$

$$\textit{Reception Energy Overhead}: \ 0.329/(1024)^{-2} \times k\left(N^2P - 3NP + 12Q + 2P\right) \text{ J}$$

Illustrative Example

Assuming $k = 4, M = 50, N = 50, P = 10, S = 5, Q = 15, n = 5$ the energy consumption for running the presented technique for the entire operational period sums up to 2736.15 J. The energy consumed for running background processes is 0.562 J/s, and initial energy of the smartphone is 35,568 J. Hence, a smartphone dedicated for running the presented technique can function for almost 13 h for the present application, i.e. for the full operational period in a day, without further charging.

5.4.4.2 Quantitative Analysis

The efficiency of the Wise-PRoPHET technique is measured through simulation.

Simulation Environment

The simulation environment used for quantitative analysis of the multilayer encryption and hashing-based scheme, described in Sect. "Simulation Environment", is used for conducting quantitative performance analysis of Wise-PRoPHET as well.

Simulation Set-up

The simulation set-up used for the simulation of multilayer encryption and hashing-based scheme, described in Sect. "Simulation Set-up", is used for Wise-PRoPHET as well with some minor modifications. The routing model adopted is PRoPHET [43]. The forwarding evidences (Sect. "Monitoring Forwarding Behaviour") are of size 48 bytes. Benign, selfish (*SN*) and malicious (*MN*) nodes are considered in the network. Benign nodes participate in routing activity following the Wise-PRoPHET technique. Selfish nodes drop all messages (directed to other nodes) forwarded to it. Malicious nodes set the direct trust value to 1 for all nodes whose direct trust comes out to be less than 0.5 using Eq. 5.1a and set direct trust value to 0 for all nodes whose direct trust comes out to be more than or equal to 0.5. The *timeout* period (time for which a node must wait before identifying its neighbour as selfish) is set as 360 s for $1/\lambda = 3600$ s. The basis of setting such value for timeout is as per the guidelines provided in Sect. 5.4.3.3. Apart from the simulation parameters stated in Table 5.3, a few more parameters are used for the simulation of Wise-PROPHET, mentioned in Table 5.10.

Table 5.10 Parameters used for simulation of Wise-PRoPHET

Parameter	Value
timeout	360 s
μ, ω_1, ω_2	[0, 1]
θ	[0, 0.1]
S	[4, 10]
τ	0.5

Simulation Metrics

Performance of Wise-PRoPHET is evaluated in two stages: measuring the extent of achieving major design goals and evaluation of network performance. Metrics used for measuring the extent of achieving major design goals are:

Estimation Bias—Bias of an estimator is the difference between the estimator's (local trust, in this case) expected value and the true value of the parameter (global trust, in this case) being estimated. Thus, estimation bias indicates how accurately the presented technique can estimate the global trust of a forwarder and is defined as

$$\text{Estimation Bias} = |E(\text{local trust}) - \text{global trust}|$$

where E(local trust) denotes the expectation of local trust, defined in Eq. 5.2c.

Detection Ratio—It indicates how well the presented technique can identify selfish nodes [3]. We consider two types of such ratios:

$$\text{Ratio of True Positive} = \frac{\text{No. of correct detections}}{\text{No. of selfish nodes in the network}}$$

$$\text{Ratio of False Positive} = \frac{\text{No. of false detections}}{\text{No. of nodes detected as selfish}}$$

Attraction Ratio—It indicates how well the technique can restrict forwarding messages to selfish nodes [2] defined as

$$\text{Attraction Ratio} = \frac{\text{No. of msgs. received by selfish nodes}}{\text{No. of msgs. relayed}}$$

Metrics used for measuring network performance are: delivery ratio, average delay and overhead ratio. These metrics are as defined in Sect. "Simulation Metrics".

Finally, another metric called *tolerance level* to measure the comprehensive performance of the presented technique is used. It is defined as the maximum percentage of selfish and nodes that can be tolerated for achieving a given level of network performance, in terms of delivery ratio, average delay and overhead ratio.

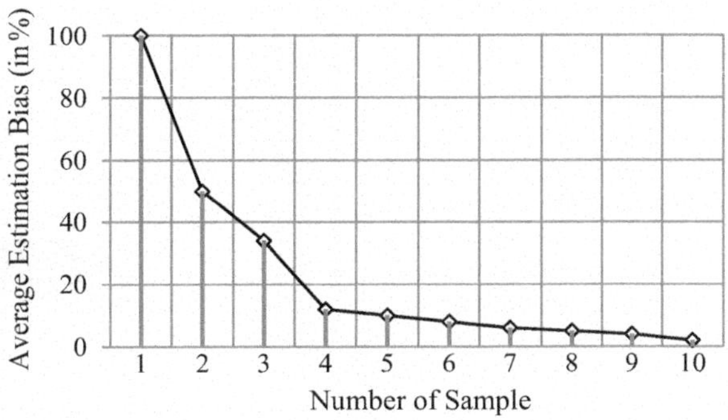

Fig. 5.13 Bias in estimating global trust against sample size

Results and Discussion

The presented technique is contrasted with two other competing techniques T-threshold [3] and T-PRoPHET [1], in terms of both design goals and network performance. T-PRoPHET does not deal with bad-mouthing and ballot-stuffing, and hence the percentage of malicious nodes in T-PRoPHET is supposed to be 0. The results presented here are the average of 50 independent runs.

– ***Achieving Major Design Goals***: Three sets of experiments are conducted for evaluating the performance of the present technique by measuring the extent of attaining design objectives.

As mentioned in Sect. "Estimating Global Trusts", the accuracy of the estimated global trust depends upon the number of samples (S, in this case) being considered for estimation. Hence, in the first set of experiment, the average estimation bias is measured across all nodes against the number of samples to figure out the optimum number of samples to be considered. A fixed (50%) number of selfish and malicious nodes are considered to exist in the network. As expected, Fig. 5.13 exhibits that the average estimation bias lessens with increasing sample size. It is observed that a sample size of 4 lessens the estimation bias to less than 10%. Accordingly, all further experiments are conducted by varying the sample size from 4 to 10.

The speed, at which a forwarder-node generates the global perception about the forwarding behaviour of its neighbours, has a direct impact on the identification of selfish nodes. Thus, in the second set of experiment, the timeliness of global trust generation at each forwarder-node is measured. The simulation results on estimation bias, obtained in the first set of experiment, are used to compute the global trust accuracy at each node. Figure 5.14 depicts the average global trust accuracy across all nodes against time, with a fixed (50%) number of selfish nodes. It is observed that the average global trust accuracy reaches 90% after 7 h of simulation in the presence of 50% malicious nodes, Fig. 5.14a; reaches 90% after 6 h of simulation in the presence of 25% malicious nodes, Fig. 5.14b; and reaches 90% after 4 h of simulation in the absence of malicious nodes, Fig. 5.14c.

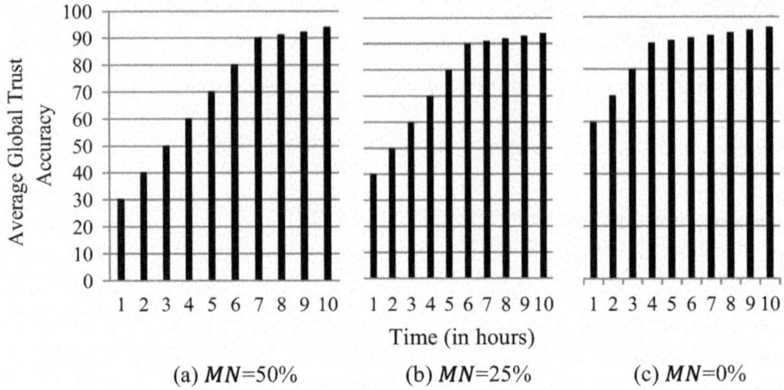

Fig. 5.14 Average global trust accuracy over time

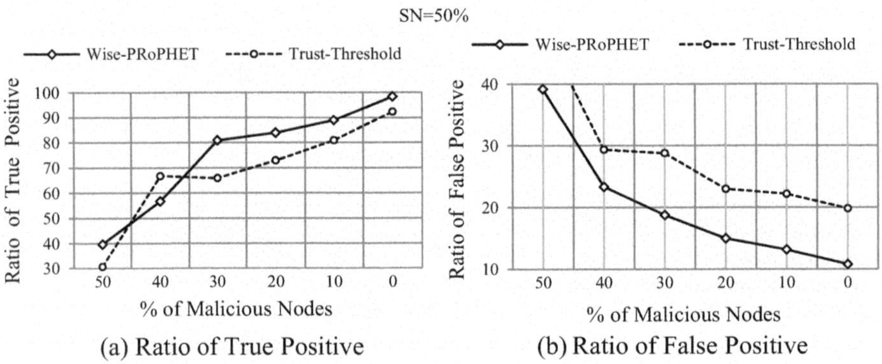

Fig. 5.15 Detection ratio with different levels of maliciousness

In the third set of experiment, variation in detection ratios with different levels of maliciousness is measured. These ratios are directly linked with bad-mouthing and ballot-stuffing and hence vary with the percentage of malicious nodes. Figure 5.15 plots detection ratios with varying number of malicious nodes and a fixed (50%) number of selfish nodes. It is observed from the figure that detection ratio improves with reducing number of malicious nodes in both the techniques. Improvement rate in the presented technique is higher compared to trust threshold. Precisely, Fig. 5.15a indicates that Wise-PRoPHET outperforms trust threshold beyond 35% malicious nodes and has an average of 10% more true identifications. Figure 5.15b depicts that Wise-PRoPHET performs consistently better and has an average of 12% less false identifications than trust threshold.

The fourth set of experiment measures attraction ratio for three competing techniques, assuming 50% malicious nodes. Figure 5.16 depicts that all three techniques show a reducing trend, but T-PRoPHET performs best owing to the absence of malicious nodes. Figure 5.16a exhibiting the variation in attraction ratio against time

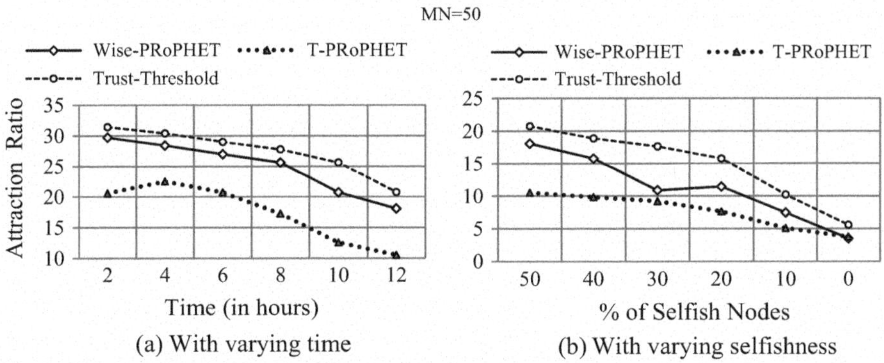

Fig. 5.16 Attraction ratio varying with time and selfishness

depicts that Wise-PRoPHET has an average of 7% better performance than trust threshold. Figure 5.16b showing the variation in this ratio against percentage of selfish nodes justifies that Wise-PRoPHET has an average of 9% better performance than trust threshold.

– ***Evaluation of Network Performance***: To measure network performance while achieving the design goals, five sets of experiments are conducted.

As mentioned in Sect. 5.4.2.3, the number of messages delivered and its speed through the Wise-PRoPHET technique depends on the value of the self-trust rationalizing factor (α). Hence, in the first set of experiment, we attempt to figure out the optimum value of α that should be considered for conducting further experiments on delivery ratio and average delay. Figure 5.17a, b plots delivery ratio and average delay, respectively, of Wise-PRoPHET against different values of α. Average of all the results obtained for different percentages of selfish and malicious nodes is plotted. On the one hand, it is observed that, for lesser values of α, delivery ratio lessens due to the inclusion of selfish nodes resulting in frequent packet drops. On the other hand, for bigger values of α, the ratio lessens due to the effect of self-trusting. Also, average delay increases consistently with increasing value of α, as nodes with high delivery predictability and/or high forwarder trust are chosen for forwarding messages. Delivery ratio is the highest for $\alpha = 0.7$. Hence, we conduct all further experiments with $\alpha = 0.7$.

In the second set of experiment, the performance of Wise-PRoPHET is compared with other competing schemes in terms of delivery ratio. T-PRoPHET performs the best, as the percentage of malicious nodes in T-PRoPHET is 0. However, T-PRoPHET beats Wise-PRoPHET by only 3.25% in the presence of 50% malicious nodes, Fig. 5.18a. It beats by only 0.18% in the presence of 25% malicious nodes, Fig. 5.18b. Wise-PRoPHET beats T-PRoPHET by 25.5% and trust threshold by 23% in the absence of malicious nodes, Fig. 5.18c. Thus, Wise-PRoPHET performs almost as good as T-PRoPHET in the presence of malicious nodes and outperforms it in the absence of such nodes. This is because T-PRoPHET suffers from

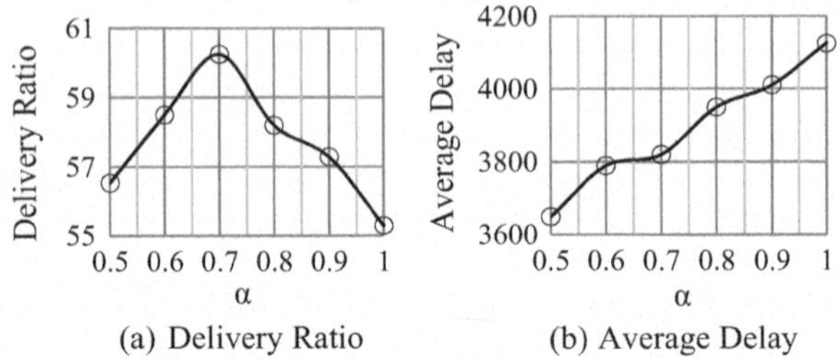

Fig. 5.17 Impact of self-trust rationalizing factor

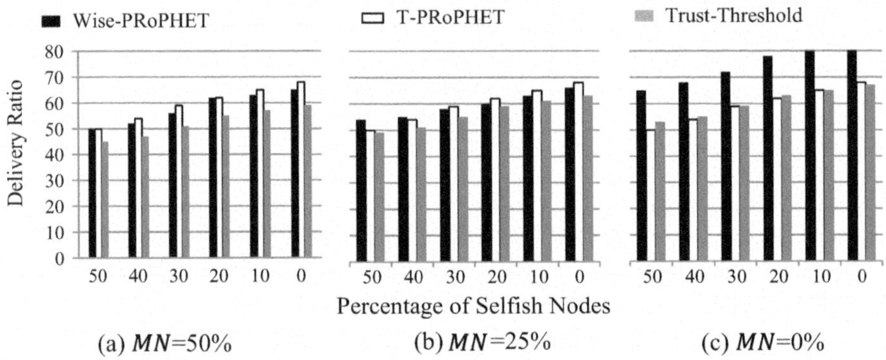

Fig. 5.18 Delivery ratio with different levels of selfishness

the self-trusting effect and Wise-PRoPHET reduces this effect by choosing $\alpha = 0.7$. Moreover, Wise-PRoPHET, unlike others, uses global trust values, for computing forwarding competencies, instead of only direct trust values.

In the third set of experiment, the performance of Wise-PRoPHET is compared with other schemes in terms of average delay. Trust threshold performs the best. Nevertheless, the average delay is only 8.8% less than Wise-PRoPHET in the presence of 50% malicious nodes, shown in Fig. 5.19a; 6.7% less in the presence of 25% malicious nodes, shown in Fig. 5.19b, and 2.6% less in the absence of malicious nodes, shown in Fig. 5.19c. The reason for this delay in Wise-PRoPHET is the message delay incurred due to the forwarding evidences, but then these evidences make Wise-PRoPHET better than trust threshold in terms of detection ratio, attraction ratio and delivery ratio. But, Wise-PRoPHET outperforms T-PRoPHET owing to the use of $\alpha = 0.7$.

The fourth set of experiment measures overhead ratio of all the three techniques. For Wise-PRoPHET, control packets refer to the forwarding evidences, for T-PRoPHET they refer to positive feedback messages, and for trust threshold these are the different types of tokens. Figure 5.19 depicts that for the Wise-PRoPHET technique *FE* packets can be at most 14% of all transmitted packets. Multiple *FE*s directed to a specific node are clubbed together to reduce the overhead. T-PRoPHET outperforms Wise-PRoPHET by 4% in case of 50% malicious nodes, Fig. 5.20a, and by 3% in case of 25% malicious nodes, Fig. 5.20b. Nonetheless, both Wise-

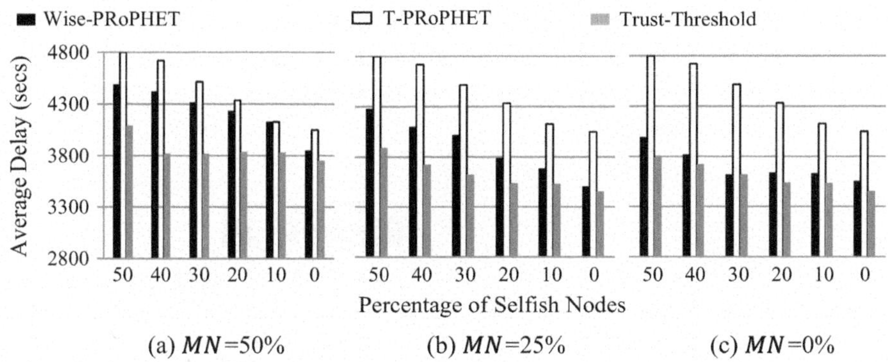

Fig. 5.19 Average delay with different levels of selfishness

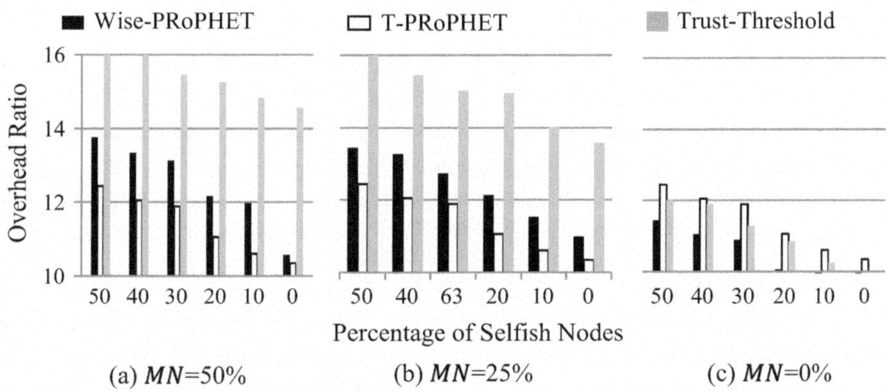

Fig. 5.20 Overhead ratio with different levels of selfishness

PRoPHET (by 18%) and trust threshold (by 5%) beat T-PRoPHET in the absence of malicious nodes, Fig. 5.20c.

Figure 5.21 plots delivery ratio of Wise-PRoPHET with varying percentage of selfish and malicious nodes. The plot provides us a design guideline to decide the maximum percentage of selfish and malicious nodes that can be tolerated to achieve a given level of delivery ratio. It is observed that our technique tolerates up to 30% selfish nodes and 40% malicious nodes to achieve a delivery ratio of 65%. Similar design guidelines can be drawn for achieving the desired level of average delay and overhead ratio.

In this work, a trust-based Watchdog technique that monitors its neighbouring nodes to generate a local perception about their forwarding behaviour is presented. An absorbing finite-state Markov chain is used to determine the expected message delays that assist a node in accurately determining the forwarding behaviour of its neighbours. This local knowledge is then shared with other nodes aiming at generating a global perception for identifying selfish nodes in the network. The Watchdog

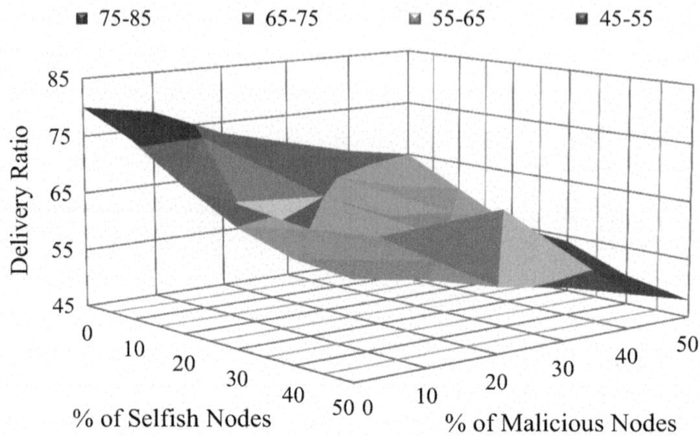

Fig. 5.21 Delivery ratio of Wise-PRoPHET with different levels of selfishness and maliciousness

also identifies the malicious nodes that collude to either spoil the reputation of a benign node or boost the reputation of a selfish node. Finally, the Watchdog is flawlessly incorporated with PRoPHET routing protocol for on-the-fly identification and avoidance of selfish and malicious forwarders in a DTN-leveraged post-disaster communication network. The integration successfully eliminates self-trusting, a property of trust-based data forwarding in opportunistic networks which reduces and delays message transfers. Overhead analyses in terms of computation and communication requirements are done to justify the application of Wise-PRoPHET in a resource-limited low-power smartphone-based DTN. Simulation results validate that the technique does better than a couple of state-of-the-art contending schemes. Wise-PRoPHET provides a design guideline to decide the percentage of selfish and malicious nodes that can be tolerated to achieve a given delivery ratio. In particular, it tolerates 30% selfish and 40% malicious nodes to deliver up to 60% situational messages in the present application domain.

5.5 Conclusion

In this chapter, three security solutions are presented to encounter the stringent security requirements of a DTN. The protocols ensure accuracy and timeliness of resource needs at the control station, enabling rational and appropriate resource planning. First, a multilayer encryption and hashing-based scheme is presented to protect the transmitted emergency resource needs against privacy attacks like authenticity, eavesdropping and modification. Next, a security-aided and group encounter-based PRoPHET routing protocol is presented for secured dissemination of categorized resource needs from different shelters to the relief camps. The protocol provides security against privacy attack like eavesdropping and communication attacks like black hole, identity spoofing and bundle store overflow. Finally, a trust-based Watchdog technique is presented and seamlessly integrated with PRoPHET to provide security against communication attack like packet drop and trust/reputation attacks like bad-mouthing and ballot-stuffing. These security solutions run on the smartphones, carried by volunteers, and contribute towards reliable data dissemination so that resource planning services can be rendered reliably.

References

1. Li N, Das S (2013) A trust-based framework for data forwarding in opportunistic networks. Ad Hoc Netw 11(4):1497–1509
2. Dias JAFF, Rodrigues JPC, Xia F, Mavromoustakis CX (2015) A cooperative watchdog system to detect misbehavior nodes in vehicular delay-tolerant networks. IEEE Trans Ind Electron 62(12):7929–7937
3. Chen IR, Bao F, Chang M, Cho JH (2013) Dynamic trust management for DTN and its application in secure routing. IEEE Trans Parallel Distrib Syst 25(5):1200–1210
4. Zhu H (2009) Security in DTNs. Ph.D. Thesis, Electrical and Computer Engineering, University of Waterloo. Available at https://uwspace.uwaterloo.ca/bitstream/handle/10012/4348/Haojin%20Zhu's%20Phd%20Thesis.pdf. Accessed July 2017
5. Durst RC (2002) An infrastructure security model for delay tolerant networks. Available at www.dtnrg.org/papers/dtn-sec-wp-v5.pdf. Accessed July 2017
6. Asokan N et al (2007) Towards securing disruption-tolerant networking. Nokia Research Centre, NRC-TR-2007-007
7. Defrawy KE, Solis J, Tsudik G (2009) Leveraging social contacts for message confidentiality in delay tolerant networks. Proceedings of COMPSAC 2009:271–279
8. Goldschlag DM, Reed MG, Syverson PF (1996) Hiding routing information. In: Proceedings of workshop on information hiding, pp 137–150
9. Shikfa A, Onen M, Molva R (2009) Privacy in content-based opportunistic networks. Proc WAINA 2009:832–837
10. Sakai K (2016) An analysis of onion-based anonymous routing for delay tolerant networks. In: Proceedings of ICDCS, 2016
11. Edelman PT, Donahoo MJ, Sturgil DBl (2010) Secure group communications for DTNs. In: Proceedings of ICITST 2010
12. Le Z, Vakde G, Wright M (2009) PEON: privacy-enhanced opportunistic networks with applications in assistive environments. In: Proceedings of PETRA 2009, Article No. 76
13. Gupta AK et al (2017) A review on secure data retrieval using AES approach for decentralized disruption-tolerant military network. Int J Innov Res Comput Commun Eng 5(3):6285–6289
14. Cabaniss R, Kumar V, Madria S (2015) Multi-party encryption (MPE): secure communications in delay tolerant networks. Wireless Netw 21(4):1243–1258
15. Zhu H et al (2014) A probabilistic misbehavior detection scheme towards efficient trust establishment in DTN. IEEE Trans Parallel Distrib Syst 25(1):22–32
16. Chen IR, Bao F, Chang M, Cho JH (2011) Integrated social and QoS trust-based routing in delay tolerant networks. Wireless Pers Commun 66(2):443–459
17. Dini G, Duca AL (2012) Towards a reputation-based routing protocol to contrast blackholes in a delay tolerant network. Ad Hoc Netw 10(7):1167–1178
18. Ayday E, Lee H, Fekri F (2010) Trust management and adversary detection for DTN. In: Proceedings of MILCOM 2010
19. Jethawa H, Madria S (2017) Incentive mechanism for data-centric message delivery in delay tolerant networks. In: Proceedings of ICDCS 2017, pp 2551–2554

20. Boneh D, Crescenzo G, Ostrovsky R, Persiano G (2004) Public-key encryption with keyword search. In: Proceedings of Eurocrypt 2004, pp 506–522
21. Hsu ST, Yang CC, Hwang MS (2013) A study of public key encryption with keyword search. Int J Netw Secur 15(2):71–79
22. Shikfa A, Önen M, Molva R (2009) Privacy in context-based and epidemic forwarding. In: Proceedings of WOWMOM 2009, pp 1–7
23. Abdalla M et al (2005) Searchable encryption revisited: consistency properties, relation to anonymous IBE, and extensions. In: Proceedings of advances in cryptology—CRYPTO 2005, pp 205–222
24. Waters BR, Balfanz D, Durfee G, Smetters DK (2004) Building an encrypted and searchable audit log. In: Proceedings of NDSS symposium 2004, pp 5–6
25. Golle P, Staddon J, Waters B (2004) Secure conjunctive keyword search over encrypted data. In: Proceedings of ACNS 2004, pp 31–45
26. Basu S, Roy S (2014) A group-based multilayer encryption scheme for secure dissemination of post-disaster situational data using peer-to-peer delay tolerant network. In: Proceedings of IEEE ICACCI 2014, pp 1566–1572
27. Basu S, Roy S, DasBit S, Bandyopadhyay S (2016) A human mobility based knowledge sharing approach for post disaster need assessment using DTN. In: Proceedings of ACM ICDCN 2016, Article 34
28. Emergency Response Units (ERUs): Types. International Federation of Red Cross and Red Crescent Societies. Available at http://www.ifrc.org/en/what-we-do/disastermanagement/responding/disaster-response-system/dr-tools-and-system/eru/types-of-eru/. Accessed May 2015
29. Emergency Disaster Relief (2015) Japan International Corporation Agency. Available at http://www.jica.go.jp/english/our_work/types_of_assistance/emergency.html. Accessed May 2015
30. Bhattacharjee S, Basu S, Roy S, DasBit S (2016) Best-effort delivery of emergency messages in post-disaster scenario with content-based filtering and priority-enhanced PRoPHET over DTN. In: Proceedings of IEEE COMSNETS 2016, pp 1–7
31. Bhattacharjee S, Roy S, Bandyopadhyay S (2014) Exploring an energy-efficient DTN framework supporting disaster management services in post disaster relief operation. Wireless Netw 21(3):1033–1046
32. Rifa-Pous H, Herrera-Joancomarti J (2011) Computational and energy costs of cryptographic algorithms on handheld devices. In: Proceedings of Future Internet 2011, pp 31–48
33. Potlapally N, Ravi S, Raghunathan A, Jha N (2006) A study of the energy consumption characteristics of cryptographic algorithms and security protocols. IEEE Trans Mob Comput 5(2):128–143
34. Google Map of water, food, shelter and medical resources for Nepal earthquake. Available via https://www.google.com/maps/d/viewer?mid=1Iv7GILViqyJAFn5o5hi1F2Fg8mc&hl=en_US (2015). Accessed July 2016
35. Keränen A, Ott J, Kärkkäinen T (2009) The ONE simulator for DTN protocol evaluation. In: Proceedings of SIMUTOOLS 2009, Article No. 55
36. Doctors For You. Available at http://doctorsforyou.org. Accessed July 2017
37. Benfield A (2015) Nepal earthquake event recap report. Available via http://thoughtleadership.aonbenfield.com/documents/201509-nepal-earthquake.pdf (2015). Accessed July 2016
38. Uddin MYS, Nicol DM, Abdelzaher TF, Kravets RH (2009) A post-disaster mobility model for delay tolerant networking. In: Proceedings of Winter Simulation Conference 2009, pp 2785–2796
39. Spyropoulos T, Psounis K, Raghavendra CS (2005) Spray and wait: an efficient routing scheme for intermittently connected mobile networks. In: Proceedings of SIGCOMM workshop on delay-tolerant networking 2005, pp 252–259
40. Wang B, Chen X, Chang W (2014) A light-weight trust-based QoS routing algorithm for AdHoc networks. Pervasive Mob Comput 13:164–180
41. Basu S, Bhattacharjee S, Roy S, Bandyopadhyay S (2015) SAGE-PRoPHET: a security aided and group encounter based PRoPHET routing protocol for dissemination of post disaster situational data. In: Proceedings of ACM ICDCN 2015, Article No. 20

42. Basu S, Roy S (2014) Secured categorization and group encounter based dissemination of post disaster situational data using peer-to-peer delay tolerant network. In: Proceedings of FICTA 2014, Springer AISC J 328(2):747–755
43. Lindgren A, Doria A, Schelén O (2003) Probabilistic routing in intermittently connected networks. SIGMOBILE Mob Comput Commun Rev 7(3):19–20
44. Uttarakhand Disaster Report. NDTV. Available via http://www.ndtv.com/article/india/uttarakhand-1-350-evacuated-from-badrinath-united-nations-says-over-11-000-issing386683. Accessed May 2015
45. Uttarakhand: thousands still stranded, see over-view. NDTV. Available via http://www.ndtv.com/article/india/uttarakhand-thousands-still-stranded-see-over-view-383781. Accessed May 2015
46. Grasic S, Davies E, Lindgren A, Doria A. The evolution of a DTN routing protocol—PRoPHETv2. In: Proceedings of CHANTS 2011, pp 27–30
47. Burgess J, Gallagher B, Jensen D, Levine BN (2006) MaxProp: routing for vehicle-based disruption-tolerant networks. In: Proceedings of INFOCOM 2006, pp 1–11
48. Vahdat A, Becker D (2000) Epidemic routing for partially-connected Ad Hoc networks. Master's Thesis, Duke University, Durham, NC, USA
49. Basu S, Biswas A, Roy S, DasBit S (2018) Wise-PRoPHET: a watchdog supervised PRoPHET for reliable dissemination of post disaster situational information over smartphone based DTN. J Netw Comput Appl 109(2018):11–23
50. Orallo EH et al (2015) CoCoWa: a collaborative contact-based watchdog for detecting selfish nodes. IEEE Trans Mob Comput 14(6):1162–1175
51. Marti S, Giuli TJ, Lai K, Baker M (2000) Mitigating routing misbehaviours in MANETs. Proceedings of MobiCom 2000:255–265
52. Marmol FG, Perez GM (2012) TRIP, a trust and reputation infrastructure-based proposal for vehicular Ad Hoc networks. J Netw Comput Appl 35(3):934–941
53. Zhou P et al (2015) Toward energy-efficient trust system through watchdog optimization for WSNs. IEEE Trans Inf Forensics Secur 10(3):613–625
54. Cho Y, Qu G, Wu Y (2012) Insider threats against trust mechanism with watchdog and defending approaches in WSNs. In: Proceedings of SPW 2012, pp 134–141
55. Lehmann EL, Casella G (1998) Theory of point estimation. Springer Texts in Statistics
56. Groenevelt R, Nain P, Koole G (2005) The message delay in mobile Adhoc networks. Perform Eval 62(1–4):210–228
57. Haas Z, Small T (2006) A new networking model for biological applications of Ad Hoc sensor networks. IEEE/ACM Trans Netw 14(1):27–40
58. Guo XF, Chan MC (2013) Plankton: an efficient DTN routing algorithm. In: Proceedings of SECON 2013, pp 494–502
59. Wang W, Zwolinski M (2013) An improved instruction-level power model for ARM11 microprocessor. In: Proceedings of HIP3ES 2013, pp 1–7

Chapter 6
Conclusion

The availability of cellular and Internet connectivity in a post-disaster scenario is ruled out more often than not due to the failure of the supporting infrastructure through both system damage and system overuse. Therefore, the possibility of exchanging situational information using normal communication infrastructure gets almost prohibited resulting in high perception deviation at the control station about the actual needs of the remote shelters. Under such circumstances, providing post-disaster services like *need assessment*, *need accumulation*, *resource planning* and *reliable data dissemination* becomes truly challenging leading to futile disaster relief endeavours. The works in this book deal with these challenges and devises strategies to offer a framework for reliable post-disaster services over smartphone-based DTN.

This concluding chapter imparts a review of the research described in this book and highlights important contributions of this work. It also discusses the possible directions for future work in the area of rendering reliable post-disaster services in the absence of traditional communication facilities, using alternative solutions. Finally, it provides concluding remarks.

6.1 Summary of the Proposed Framework

This section summarizes the works done in this book for accurate need assessment, timely need accumulation, efficient resource planning and reliable data dissemination over a smartphone-based DTN.

In Chap. 2, a principal component regression (PCR) model is designed for targeting forecast of post-disaster emergency resource demands based on the dynamically changing situational parameters. The process of deriving the model comprises three distinct steps—(i) identification of influencing parameters through literature survey, (ii) selection of uncorrelated influencing parameters by means of principal component analysis, and (iii) formulation of the forecasting model using multiple linear regression on principal components. Experimental results justify that the model projects

S. Basu et al., *Reliable Post Disaster Services over Smartphone Based DTN*, Smart Innovation, Systems and Technologies 137,
https://doi.org/10.1007/978-981-13-6573-7_6

resource demands in the shelters with noticeable accuracy (accuracy factor nearly 1). Current values of the situational parameters of a shelter like—number of victims, the type and degree of emergency they have, their consumption pattern, rate of resource replenishment and rate of migration are uploaded to the shelter-node (a laptop or workstation in the shelter), installed therein, by the emergency managers present on site. The shelter-node, on the basis of these situational parameter values, forecasts the future demand for different emergency resources using the proposed PCR model.

While responding to a post-disaster emergency situation with constrained availability of resources, failing to understand the utilities of those resources leads to futile allocation strategies. Utility of an emergency resource is its importance at a particular instance of time. While demand represents the quantity of resource required, utility depicts the urgency of that requirement. However, quantifying a subjective notion-like utility is somewhat challenging. Therefore, in the next work in Chap. 2, a utility function is derived and used for enumerating the utility of different emergency resources at the shelters. The function expresses utility of an emergency resource in terms of the amount of that resource allocated at different time points. It is observed that importance of a resource in a shelter decreases with time if the resource is periodically allocated to that shelter; it increases if not allocated after a certain time, leading to starvation for that resource. As in case of the PCR model, current values of the situational parameters of a shelter are uploaded to the shelter-node by the emergency managers present on site. The shelter-node enumerates the utility of such resources using the derived utility function. The demand forecast model and the utility function assist in accurate assessment for emergency resource needs at different shelters.

The demand forecast model and the utility function, derived in Chap. 2, are used to assess the exact need (i.e. demands and utilities) for emergency resources at different shelters. These assessed needs are required to be transmitted and accumulated at the control station for maintaining a fully functional relief chain. In the absence of stable communication infrastructure during disasters, it becomes difficult to collect the resource needs from remote and inaccessible shelters and transmit them to the control station. All these result in high perception deviation at the control station about the actual needs, leading to inappropriate resource planning. This calls for designing a need accumulation scheme that would collect and transmit the shelter resource needs to the control station, even in the absence of stable communication infrastructure. In Chap. 3, assuming an underlying DTN-leveraged-smartphone-based communication network (carried by volunteers and relief workers), an opportunistic knowledge sharing scheme is proposed. The scheme uses human mobility characteristics to transmit and accumulate emergency resource needs of far-flung shelters to the control station in the absence of stable communication infrastructure. In this scheme, a volunteer periodically captures knowledge about resource needs (demand and utility of resources) in different shelters, broadcast by the shelter-node therein; opportunistically exchanges this knowledge with other volunteers whom she meets due to mobility; and creates a global view of resource needs across all shelters in the entire disaster-affected area. Such exchanges are implemented with the help of Bluetooth or Wi-Fi Direct interface of the smart phones following DTN routing protocols. When any one of the forwarder-nodes manages to visit the control station (located outside the affected area), it delivers its knowledge about the global view of resource needs therein. The proposed scheme requires only one volunteer to visit the control station to deliver a near accurate view of the unified resource needs of the entire area instead of requiring one volunteer from each shelter carrying the resource needs of that shelter. Simulation results justify that the scheme succeeds to transmit resource needs to the control station even in a sparse network environment within the operational period. Although such need information may not be accurate to the highest level of precision but can always serve as a basis for resource allocation to a far-flung shelter whose requirements were otherwise unknown. Thus, the opportunistic knowledge sharing scheme assists in successful accumulation of emergency resource needs at the control station which are then used for subsequent resource planning.

The opportunistic knowledge sharing scheme, proposed in Chap. 3, is used to accumulate the emergency resource needs at the control station, over a smartphone-based DTN, for necessary resource planning. Now, DTNs are characterized by inconsistent end-to-end paths, intermittent connectivity and unpredictable latency leading to dropping and distortion of transmitted messages. Therefore, the possibility of receiving corrupted or stale information, or even no information, at the control station cannot be ruled out. Such events lead to inaccurate and partial knowledge about the emergency resource needs at the shelters. Therefore, the opportunistically received resource needs are required to be thoroughly verified before they are used for resource allocation. Furthermore, resources in a post-disaster scenario are scarce and not all demands can be met. Thus, minimizing the deficit in allocation of high-utility resources is crucial. On the other hand, since the underlying communication environment uses DTN, information about resource needs at the shelters reach significantly late at the control station. Hence, minimizing the resource deployment time is inevitable. In Chap. 4, a case-based reasoning (CBR) leveraged need validation technique is formulated to validate the PCR forecast demands that are opportunistically transmitted to the control station over a smartphone-based DTN, before they are used for emergency resource planning. Experimental results justify that the technique project resource needs in the shelters with noticeable accuracy (accuracy factor nearly 1). The technique, executed at the control station, (a) verifies the received resource needs, (b) rectifies the corrupted or stale needs and (c) projects the unavailable needs. The proposed technique enables the control station to carry out effective resource allocation.

After validating the received resource needs, the next task is to formulate an optimal resource allocation model. For this, Chap. 4 proposes a utility-based integer programming model for optimal resource allocation using the opportunistically transmitted and CBR-validated resource needs. The formulated model minimizes the overall resource deficit of high-utility resources at different shelters and also minimizes the total resource deployment time to achieve fast and effective disaster relief. The proposed model reduces deficit of emergency resources at a much faster rate in shelters that have higher utility for them, than in shelters that have comparatively lower utility. The novelty of the model lies in the fact that it is utility driven and works with impaired network connectivity. Simulation results justify that proposed model performs better than other competing schemes in a fully connected network environment and performs almost as good as other schemes even in a sparse network environment. The control node (a workstation or a server at the control station) receives updated resource needs from the forwarder-nodes; carries out a case-based reasoning driven technique to validate/estimate the needs; and executes a utility driven optimal resource allocation strategy to allocate resources to shelters targeting minimization of resource deficit and deployment time. The CBR driven need validation technique and the integer programming model for optimal resource allocation brings about efficient resource planning.

The success of effective resource planning depends upon the availability of reliable situational data, i.e., shelter resource needs, at the control station. These resource needs are transmitted using the opportunistic knowledge sharing algorithm, explained in Chap. 3 that assumes a smartphone-based DTN as the underlying communication network. Network performance and reliability of the transmitted situational data in DTNs depend upon the level of co-operation among participating nodes. As a result, DTNs are vulnerable to a number of security threats arising out of node selfishness as well as malicious intentions, posing serious threats on the accuracy and timeliness of such needs. This leads to delayed and incorrect resource planning. The unique characteristics of DTNs including: long delay, lack of end-to-end connectivity, fragmentation, opportunistic routing, limited computational and storage capability, make the existing security protocols designed for the conventional networks unsuitable for DTNs. Moreover, in case of a challenged network scenario like large-scale natural disaster, there may be only a few isolated small pockets (areas) where long distance connectivity (cellular or long distance Wi-Fi) is still available. In such scenarios, trusted third parties and public key infrastructure may cease to operate. All these make multiple-round key agreement, mutual authentication, distribution and revocation of public key certificates, verification of cryptographic signatures, etc. extremely challenging.

In this regard, Chap. 5 proposes three security solutions to suit the requirements of the underlying smartphone-based DTN leveraged post-disaster communication network. These solutions are decentralized, lightweight and robust. First, a multilayer encryption and hashing-based scheme is proposed to protect the transmitted emergency resource needs against privacy attacks like—authenticity, eavesdropping and modification. Simulation results depict that the scheme offers higher delivery ratio, lower overhead ratio and lower attraction ratio with only a negligible increase in average delay. Next, SAGE-PRoPHET, a security-aided and group-encounter-based PRoPHET routing protocol is proposed for secured dissemination of categorized resource needs from different shelters to the relief camps. Categorization is done by the forwarder-nodes, on-the-fly, without compromising on their confidentiality to expedite the delivery of such categorized needs to their appropriate destinations. Message forwarding is accomplished on the basis of encounters among volunteers of different relief groups. Security features are incorporated into the DTN routing protocol PRoPHET to provide security against privacy attack like eavesdropping and communication attacks like—black hole, identity spoofing and bundle store overflow. Simulation results depict that the protocol offers high delivery ratio, low overhead and attraction ratios with only negligible increase in average delay. Energy consumption of SAGE-PRoPHET is almost similar to PRoPHET and delivers maximum number of group messages to the respective camps. Finally, Wise-PRoPHET, a trust-based Watchdog technique is proposed and seamlessly integrated with PRoPHET to provide security against communication attacks like—packet drop and trust/reputation attacks like—bad-mouthing and ballot-stuffing. The Watchdog supervised PRoPHET routing protocol ensures successful transmission of crucial post-disaster situational information to the control station even in the presence of selfish and malicious nodes. These security solutions run on the smartphones, carried by volunteers, and contribute towards reliable data dissemination so that resource planning services can be rendered reliably. Simulation results justify that the technique outperforms a couple of state-of-the-art competing schemes. Wise-PRoPHET provides a design guideline to decide the percentage of selfish and malicious nodes that can be tolerated to achieve a given delivery ratio. These security solutions run on the smartphones, carried by volunteers, and contribute towards reliable data dissemination so that resource planning services can be rendered reliably.

6.2 Future Research Directions

This section gives an insight into the possible future works that can be carried out by augmenting the works described in the book.

- The proposed PCR-based forecasting system may be improved by using canonical correlation analysis to capture the relationship between a set of independent variables and more than one dependent variable. Demand for several emergency resources can then be simultaneously forecast without having to repeat PCR for each resource.
- As an extension to the present work, a site selection algorithm may be proposed, using the facility—location model, for optimal placement of resource warehouses in a disaster-affected area to further minimize resource deployment time under damaged road condition and restricted transport facilities.
- As future work, a distributed resource allocation model may be proposed so that nearby shelters can pool each other's resources in a decentralized manner until resources from the control station arrive.
- The proposed security solutions may made resilient attacks to node compromise attacks, where an unauthorized user physically captures a node and maliciously use the pins stored in it to claim authenticity. Time-based keys can be used for revoking the rights of compromised nodes.
- The proposed Wise-PRoPHET technique may be made more rational by considering the fact that a node has indeed forwarded a message but the corresponding forwarding evidence has not reached within the timeout period, due to the fragile network characteristics such as network partitions.
- Finally, the proposed SAGE-PRoPHET and Wise-PRoPHET techniques may be tuned to work on top of the proposed opportunistic knowledge sharing scheme to

make the scheme resilient to black hole, identity spoofing, bundle store overflow, packet drop, bad-mouthing and ballot-stuffing attacks.

6.3 Concluding Remarks

This book proposes a framework for reliable post-disaster services over smartphone-based DTN, which can operate despite the networking challenges that a disaster situation creates. The book explores a number of different aspects of this challenge, ranging from accurate need assessment, to timely need accumulation, to efficient resource allocation and reliable data dissemination. It also demonstrates the significance of the proposed schemes to this unique and challenging problem area. The book identifies a number of interesting future research directions, while still presenting a platform which already has high applicability.

Index

S. Basu et al., *Reliable Post Disaster Services over Smartphone based DTN*, Smart Innovation, Systems and Technologies 137,
https://doi.org/10.1007/978-981-13-6573-7

The manufacturer's authorised representative in the EU is Springer Nature Customer Service Centre GmbH, Europaplatz 3, 69115 Heidelberg, Germany. If you have any concerns regarding our products, please contact ProductSafety@springernature.com

Printed and bound by CPI Group (UK) Ltd, Croydon, CR0 4YY
28/07/2026
02179159-0016